HIGHTECH WRITING

HIGHTECH WRITING

HOW TO WRITE
FOR THE ELECTRONICS INDUSTRY

PAULA BELL

A Wiley-Interscience Publication

JOHN WILEY & SONS

New York / Chichester / Brisbane / Toronto / Singapore

Library of Congress Cataloging in Publication Data:

Bell, Paula.
  Hightech writing.

  "A Wiley-Interscience publication."
  Bibliography: p.
  Includes index.
  1. Technical writing. I. Title.

T11.B397 1985   808'.0666   85-9498
ISBN 0-471-81864-X

Printed in the United States of America

10 9 8 7 6 5 4 3 2 1

To Gāb

PREFACE

Technical writing is one of the fastest growing professions in the fastest growing industry in the world—electronics. Hightech writing is useful, well-paying work, exciting because each new assignment lets you learn about the latest twists and turns in the fast lane from experts at designing, building, and using computer technologies. It is the technical writer's job to be expert at writing.

Writing is a technical skill, its components as important to its efficient operation as the switching mechanisms are to computer logic. Neither works well unless the basic elements of the system work well. A good tech writer needs to learn the rules applicable to all good writing and the operations specific to tech writing for the electronics industry.

If you are a tech writer or thinking of entering the field from the humanities, sciences, engineering, or programming, then this book is for you.

This book is practical: it contains standards and boilerplates for organization, text, and format; it includes important points and details of each aspect of a publication; it takes note of, but is not dependent on, particular text entry or print capabilities; and the examples throughout are relevant to technical writing, not taken from literature or memos.

The glossary lists terms and acronyms used in discussing technical publications. Definition of terms and acronyms used *in* technical publications may be found in computer dictionaries, a few of which are listed in the bibliography, which suggests books that you might find helpful, captivating, or both. The bibliography entries vary from a sampling of basic grammar books through examples of books whose authors have made an art of writing about technical subjects.

The book is organized into an introduction and four parts:

PART I: *PHYSICAL COMPONENTS*—The pages of text, figures, and tables that make up a technical publication.

PART II: *LOGICAL COMPONENTS*—The techniques of grammar and rhetoric suited to all good writing and particularly to technical writing.

PART III: *OPERATIONS*—The procedures involved in gathering, organizing, writing, and editing a technical publication.

PART IV: *MAINTENANCE*—The care and feeding of the publication once it has been released to its first audience.

Do the part titles sound familiar? You commonly find them in publications dealing with hardware and/or software. The triple asterisks that begin and end this paragraph mean that there's a question or comment for you to think about in between. Because writers need to be involved in whatever technical material we're reading, I hope this technique will help to involve you in this material.

Additional notation includes: *italics* and double quotes, which are used in the standard manner; **bold** for explicit examples of a rule or technique; and square brackets ([ ]) surrounding bad examples, some of which came from my own drafts—and some of which slid through to publication.

I, of course, take responsibility for errors that slide through and credit those who have contributed to my learning about the parts of this book that help you to learn about technical writing. Thank you to San Jose State University for helping me learn while teaching, to ISS for letting me tear down a disk drive in my office, to Amdahl Corporation for guiding me through the IBM-compatible software and hardware labyrinth, to Trilogy Systems for giving me the chance to put theory into practice, and to Silicon Compilers Incorporated for the opportunity for more practice.

Thank you to Virginia Scales for scrupulous editing and constant support, to Charlotte Evans for unflinching text processing, and to Marylou Nohr for essential final touches.

Most of all, thank you to all the people who have helped me learn about the electronics industry, from those who pointed out the difference between CPU and CYA to those who explained the chemical complexities of transistor fabrication.

PAULA BELL

Morgan Hill, California
August 1985

CONTENTS

FIGURES   xvii

TABLES   xix

INTRODUCTION   1

PART I: PHYSICAL COMPONENTS   **5**

CHAPTER 1– FRONT AND BACK MATTER   7
          1.1 Front Matter   7
          1.2 Back Matter   14

CHAPTER 2– SECTIONS   17
          2.1 Page Numbers   17
          2.2 Running Feet and Heads   18
          2.3 Headers   19
          2.4 Text   21
          2.5 Warnings, Cautions, and Notes   21
          2.6 Trademarks and Logos   22
          2.7 Footnotes   23

CHAPTER 3– NOTATION   25

    3.1 Punctuation   25
    3.2 Acronyms and Abbreviations   34
    3.3 Numbers   37
    3.4 Symbols   39
    3.5 Computer Communication   39
    3.6 Spelling   46

CHAPTER 4– FIGURES   49

    4.1 Hardware Illustrations   51
    4.2 Screens   53
    4.3 Hierarchies   53
    4.4 Block Diagrams   56
    4.5 Flowcharts   56
    4.6 Photographs   61

CHAPTER 5– TABLES   63

CHAPTER 6– LISTS   67

CHAPTER 7– REFERENCES   71

    7.1 Internal   71
    7.2 External   73

CHAPTER 8– APPENDIXES   75

CHAPTER 9– GLOSSARY   77

CHAPTER 10– INDEX   79

PART II: LOGICAL COMPONENTS   **83**

CHAPTER 11– GRAMMAR   85

    11.1 Terms   86

11.2 Rules   98
11.3 Structures   104

CHAPTER 12– DICTION   123

12.1 Consistency   123
12.2 Terminology   124
12.3 Precision   125
12.4 Gender   131
12.5 Economy   132
12.6 Aesthetics   134

CHAPTER 13– DEFINITION   135

13.1 Words   136
13.2 Ways   137
13.3 Forms   140

CHAPTER 14– DESCRIPTION   143

14.1 Physical   143
14.2 Functional   145
14.3 Procedural   146

CHAPTER 15– STYLE   151

15.1 Prescriptions   151
15.2 Proscriptions   152

PART III: OPERATIONS   **155**

CHAPTER 16– PLANNING   157

16.1 Planners   157
16.2 Plans   158
16.3 Progress   164

CHAPTER 17– RESEARCH   165

17.1 Documentation   165

17.2 Interviews   167
17.3 Hands-On   172
17.4 Classes   173
17.5 Conservations   173

CHAPTER 18– WRITING   175

18.1 Organize   175
18.2 Write   177

CHAPTER 19– REVIEWS   179

19.1 Preparation   179
19.2 Distribution   180
19.3 Collection   182
19.4 Incorporation   184

CHAPTER 20– EDITS   185

20.1 Technical   186
20 2 Mechanical   190

CHAPTER 21– PRODUCTION   193

21.1 Text   193
21.2 Format   194
21.3 Components   195
21.4 Master   196

PART IV: MAINTENANCE   197

CHAPTER 22– ARCHIVES   199

22.1 Copies   200
22.2 Distribution Lists and Signoffs   201
22.3 Reader Comments and Replies   202
22.4 Memos and Notes   202

CHAPTER 23– REVISIONS   205

23.1 Revision Package   206
23.2 Revision Master   210

CHAPTER 24– REPRINTS   213

       24.1 Revision   213
       24.2 Stock   215

BIBLIOGRAPHY   217
  I.   Grammar and Format   217
  II. Writing and Style   219
 III. Reference and Research   220
 IV. Pleasure   222

GLOSSARY   223

INDEX   227

FIGURES

1-1.	Reader Comment Form	16
3-1.	Summary Card	47
4-1.	Blowout	52
4-2.	Subject	52
4-3.	Detail	54
4-4.	Screen Display	54
4-5.	Assembly Hierarchy	55
4-6.	System Block Diagram	57
4-7.	Program Flow	59
4-8.	Process Flow	60
4-9.	System Configuration	61
19-1.	General Review Form	181
19-2.	Note to Technical Reviewers	183
22-1.	Reader Comment Reply Form	203
23-1.	Revision Package Cover Page	211

TABLES

1-1.	Table of Contents	13
1-2.	Revision History for First Release	15
3-1.	Common Abbreviations	37
3-2.	Operator Symbols	40
3-3.	Command Notation	42
4-1.	Flowchart Symbols	58
5-1.	Specifications	65
5-2.	Header Standards	65
12-1.	Economy	132
20-1.	Copy/Proofreader's Marks	187
23-1.	Updated Revision History	210

The only proper attitude is to look upon a correct understanding as a triumph against the odds. We must cease to regard mis-interpretation as a mere unlucky accident. We must treat it as the normal and probable event.

—I.A. Richards

INTRODUCTION

If you are a technical writer, then who are you?

You are a person who receives input that is usually sketchy and seldom up-to-date from which you output a finished publication that will enable people who build, install, fix, or use electronics to locate and understand the information they need to get their jobs done.

Where did you come from?

If you came from an electronics, engineering, or scientific field, then you may need to learn how written English works. There are fewer rules to learn than there are to learning how a computer or nature works, and, unlike electronics, the rules vary little over time.

If you came from the humanities, journalism, or English lit, then there are vast new fields for you to roam, finding out what 2 ** 3 - 1 really means in the grand schematic of things, and learning that you can understand more than you thought you could. Electronics may not change human nature, but they will change human habits significantly.

What are you doing here?

You are learning all the time, when you interview people, while you sit at meetings, and particularly as you write about software, firmware, or hardware; disks or tapes; chips, minis, micros, or mainframes; aerospace, defense, or artificial intelligence.

You are producing a publication to help the people in your audience do their jobs.

Who is your audience?

Technical publications are intended for readers who will use the publication for an action directly involving an electronic object within the time and environment for which they use the object.

Identifying with your audience shouldn't be difficult: Did you read the whole manual about how to use all the features on your VCR before pushing any buttons? How far did you get through the manual on how to use your new washing machine or how to set your new watch? Did you experiment so successfully that you did not have to read the instructions at all? In fact, have you ever read a set of instructions or a description of one of your appliances—including your PC—other than because there was no other choice?

Most of your audience feels the same way. Help them to avoid the things that annoy you, like a publication that carefully explains how to insert a floppy on one page and expects you to be able to translate hex variables on the next; like a figure or table that makes you turn the publication sideways; like friendly, smiling stick figures standing next to unclear or incomplete text. No amount of high quality writing will make a bad product good, but bad writing can hurt a good product when it makes the product harder to use.

How do you communicate with your audience?

Writing the way you talk, as some suggest, is not the answer. When you're talking with someone, you receive constant and immediate feedback. Your audience can ask questions, look puzzled, smile, fall asleep. When you're writing, you have to depend more heavily on your skills in presenting information in close accordance with what your audience needs to know and in accordance with what you know about them.

All publications, whether to be used internal or external to the company, by sophisticated or neophyte users, need to be readable, functional, and flexible. *Readable* means that the content is logically organized and clearly presented. *Functional* means that the forms are consistent and logical so that your reader can find information easily. *Flexible* means that the publication can be reorganized and revised easily in order to remain readable and functional.

Your readers, like you, are learning—that's why they're reading the publication. Writing as you would teach is better than writing as you would talk.

Why do you want to communicate with your audience?

Technical writing offers good pay, a chance to become a better writer by daily practice, access to exciting information and people, and a comfortable—often luxurious—work environment. You want to be as good as you can be, and your books to set the industry standard.

Now what?

1. Write about as many areas of the industry as you can before—and if—you choose to specialize.

2. Read about the high technology of your choice. There are enough to choose from. Read the trade journals that pertain to your company's involvement in the industry, but be aware of what's hot in the other parts of the industry too.

3. Join a technical communications organization, a technical organization, or better yet, one of each. Even if you don't attend meetings often, the newsletters will let you know what's going on in the geographical area and the technical disciplines.

4. Learn to use a word processor. They're just electronic typewriters, bound to replace electric typewriters just as the electrics replaced manuals. If you're still infatuated with that 1969 Remington Rand, then you might want to reconsider being a technical writer.

5. Attend classes in-house if your company has them. Attend occasional seminars, conventions, and classes outside. If your company won't foot the bill, then go anyway—they're tax deductible.

6. Read this book. The best way is to read clear through in order to get familiar with the structure, the notation (**bold** for good examples and [ ] for bad), and the considerations (***the text between asterisks***). You can then use this book for reference during the various processes of writing *your* book.

PART I: PHYSICAL COMPONENTS

Technical writing, like all writing, requires basic physical structures that need not stifle creativity but should direct it.

Just as poetry, essays, and novels have standard forms, so technical writing has standard forms, because the conventionality helps you and your readers to understand the organization and to anticipate the kinds of information it contains.

Some forms are obligatory: The information must fit between covers, be ordered according to accepted patterns, be presented in the appropriate notation.

Some forms are optional: You can detail information in an appendix, expand it with references, help your reader find it by indexes.

Some forms are expected: You can illustrate information by a figure, summarize it in a table, introduce it with a list.

Still other forms are at your disposal: You can emphasize or de-emphasize information according to the level of its header, join it by punctuation, simplify it by acronyms.

With enough practice, observing the formalities becomes second nature, and you can use them to write about more interesting things.

1

FRONT AND BACK MATTER

The following components compose the front and back matter of technical publications.

1.1 FRONT MATTER

The front matter of a book includes all of the pages that precede the first part or section in the publication, including:

Page numbers
Front cover
Frontispiece
Title page
Back-of-title-page
Table of contents
Preface

1.1.1 Page Numbers

Front matter is treated as a unit, numbered in lowercase Roman numerals. Page numbers begin on the back-of-title-page with "ii" and are right and left justified at the bottom of each page.

1.1.2 Front Cover

The front cover is printed with all the information necessary for its unique reference, at minimum the title and date of publication. Usually a publication number, often including a revision level, acts as a part number for reorder by the field or by a customer.

If your company logo is displayed, then use it in accordance with the rules.

If your books are placed in binders, then an insert at the front of the notebook acts as the front cover. If you also include a spine tab, then unless the tab is wide enough to hold full words, print the manual title on the tab from the top down. Don't stack letters.

1.1.3 Frontispiece

Occasionally a frontispiece is included in the front matter; usually it's a photograph. A frontispiece is generally located on the left-hand page facing the title page. It takes a title, but it does not require a figure number.

1.1.4 Title Page

The title page is the first right-hand page. It is usually printed on the same paper stock as the text pages, but in a marketing publication it may be printed on heavier grade paper.

The title page contains the same information as the front cover: the name, date, and publication number.

1.1.5 Back-Of-Title-Page

The back-of-title-page may hold one or more of the following: a revision notice, a trademark notice, a reference to the reader comment form, a copyright notice, and a disclaimer.

1.1.5.1   Revision Notice

The initial revision notice is boilerplate, for example:

This is the first release of this publication.

For updated versions, the revision notice should provide a reason for and a summary of the changes. If your publications include a revision history page at the back of each publication, then refer the reader to it. Revision histories are discussed in Section 1.2, "Back Matter," and in Chapter 23, "Revisions."

1.1.5.2   Trademark Notice

The trademark notice is boilerplate, for example:

MICROMECCA CORPORATION and MICROMECCA SUPER-10 are registered trademarks of the Micromecca Corporation.

The use of trademarks and logos in text is discussed in Section 2.6.

1.1.5.3   Reference to Reader Comment Form

The reference to the reader comment form is boilerplate, for example:

The last page of this publication is a reader comment form. Comments are welcome and become the property of Micromecca Corporation.

The reader comment page is discussed in Section 1.2.2.

1.1.5.4   Copyright Notice

The copyright notice—except for the date—is boilerplate, for example:

Copyright 1985 by Micromecca Corporation. Printed in the United States of America. All rights reserved. No part of this publication may be reproduced in any form without prior written consent by Micromecca Corporation.

NOTE: Check the copyright notice with the company's lawyers.

1.1.5.5   Disclaimer

The disclaimer is boilerplate, for example:

Specifications are subject to change without notice.

1.1.6   Preface

The preface starts on the right-hand page facing the back-of-title-page. It is not intended to be an overview of the subject of the publication; rather it defines the publication's purpose, audience, and organization. The preface may also contain a short list of references to prerequisite or corequisite publication.

The preface outline is boilerplate, for example:

P.1 PURPOSE
P.2 AUDIENCE
P.3 ORGANIZATION
P.4 REFERENCES

1.1.6.1   Purpose

The purpose of a tech manual is to help someone understand how to install, fix, or operate a system. State clearly what the manual is to be used for and, if there is any question, what it will not cover, for example:

P.1 PURPOSE
***The Micromecca Super-10 Maintenance Guide* describes the preventive maintenance procedures that must be per-**

formed every six months. It does not cover emergency maintenance or troubleshooting procedures.

1.1.6.2 Audience

Don't list your intended readers by title only, because titles may not be the same throughout organizations that will be using your equipment or programs. For example, the DP manager of a small shop or a field service trainee might be performing preventive maintenance.

State the audience for the book in terms of the purpose you described. And if you expect a certain level of experience from the audience, then state that, for example:

P.2 AUDIENCE
This guide is intended for use by people who will be performing scheduled maintenance on the Micromecca Super-10. It assumes that the user has completed ten hours of preventive maintenance training on the equipment.

1.1.6.3 Organization

The overall organization of the publication depends on the product and the audience. Rather than just listing section titles, which your audience can find in the table of contents, provide the strategy you used to structure the book, for example:

P.3 ORGANIZATION
Procedures are discussed in the order they are to be performed and are presented as a check list. Tools necessary for each procedure are listed at the top of the procedure. A log at the end of this guide provides space for dates and results of maintenance.

You can follow the organization strategy with the list of section titles and a sentence or phrase that describes each.

1.1.6.4 References

If any publications are prerequisite or corequisite to the information in the manual, then they should be listed under "REFERENCES," for example:

P.4 REFERENCES
Micromecca Super-10 Illustrated Parts Breakdown,
1234567-301.

Section 7.2 discusses the types of external reference formats.

1.1.7 Table of Contents

The table of contents orients the reader to the structure of the manual. It starts on the right-hand page following the preface. Its title is "Table of Contents" or "Contents," and the information is listed according to parts, sections, subsections, figures, tables, appendixes, glossary, bibliography, and indexes, and the pages on which they begin.

Each entry in the table of contents should exactly mirror the titles within the publication.

If, as is often the case, time and efficient programs for indexing are not available, then include titles through the sub-section level in the table of contents. Levels are discussed in Section 2.3, "Headers."

To make it easier for your reader to find things: in a software manual, list each command in the table of contents; in a hardware manual, list each procedure.

To make it easier for you to find things, begin the table of contents with the first draft of your manual so that it acts as a dynamic outline. Computers are made for things like automatically generating tables of contents, but even if you have to do it by hand, be sure the table of contents reflects each change you make in the structure of the manual. It is invaluable as an organization check.

Table 1-1 illustrates an example table of contents.

Table 1-1.   Table of Contents

CONTENTS

SECTION 1   INTRODUCTION 1-1
    1.1   SYSTEM HIGHLIGHTS 1-3
    1.2   SYSTEM MODES 1-5
    1.3   SYSTEM REQUIREMENTS 1-8
SECTION 2   TECHNOLOGY 2-1
    2.1   CHIPS 2-2
    2.2   CHIP CARRIERS 2-5
SECTION 3   FUNCTIONAL COMPONENTS 3-1
    3.1   CENTRAL PROCESSING UNIT 3-2
        3.1.1   Instruction Unit 3-3
        3.1.2   Execution Unit 3-5
        3.1.3   Communications Unit 3-7
    3.2   MONITOR TERMINAL UNIT 3-10
        3.2.1   Frame 3-12
        3.2.2   Electronics 3-14
        3.2.3   Keyboard 3-17
    3.3   HARD DISK UNIT 3-20
        3.3.1   Controller 3-21
        3.3.2   Media 3-23

FIGURES
    1-1.   Micromecca Super-10 System 1-1
    2-1.   Super-10 Chip 2-2
    2-2.   Chip Carrier 2-5
    3-1.   System Block Diagram 3-1
    3-2.   I-Unit Block Diagram 3-3
    3-3.   E-Unit Block Diagram 3-5
    3-4.   C-Unit Block Diagram 3-7
    3-5.   Monitor Terminal Unit 3-10
    3-6.   Electronics 3-14
    3-7.   Keyboard 3-17
    3-8.   Hard Disk Unit 3-20

TABLES
    1-1.   Requirements 1-9
    3-1.   Control Keys 3-18

Can you guess the purpose and audience for a book organized like the example? A publication about a fairly expensive system, a publication for the marketeers to distribute to executives and managers, might be organized like this.

1.2 BACK MATTER

The back matter of a technical publication includes:

Revision history
Reader comment form
Back cover

1.2.1 Revision History

The first release of a publication contains a revision history page that at least carries the title of the publication, the publication number, and the date of release.

If you also include a table listing every page in the publication, then updating the publication will be easier.

Table 1-2 is an example revision history table for a first release. Table 23-1 illustrates an updated table.

1.2.2 Reader Comment Page

The reader comment form is printed on the final right-hand page in the publication. A postage-paid reply should be printed on its back. Check with the post-office for requirements concerning the print and paper weight necessary for business reply mail. If possible, perforate the form one inch from the left margin for easy removal.

The form should identify the manual and encourage reader comments. Ask questions concerning the organization and content of the manual. Provide space for answers and directions for mailing.

Table 1-2. Revision History for First Release

Revision History

Super-10 User's Guide

1234567-000, March 1985

A list of pages in this publication and each page's current revision level follows:

Page	Revision	Page	Revision
title	000	3-1	000
ii	000	3-2	000
iii	000	3-3	000
iv	000	3-4	000
1-1	000	3-5	000
1-2	000	3-6	000
1-3	000	3-7	000
1-4	000	3-8	000
2-1	000	3-9	000
2-2	000	3-10	000
2-3	000	3-11	000
2-4	000	3-12	000
2-5	000	4-1	000
2-6	000	4-2	000
2-7	000	4-3	000
2-8	000		

Responses to reader comments are discussed in Part IV, "Maintenance."

Figure 1-1 illustrates an example reader comment form.

1.2.3 Back Cover

The back cover, whether the back of a binder or made of cardboard, should hold the company name, address, and telephone number. The company logo is optional.

Reader's Comments

Super-10 Operator's Guide
1234567-000, March 1985

Please answer the questions below and add any suggestions for improving this publication. Is the information:

	Yes	No
Adequate to the subject?	___	___
Well organized?	___	___
Clearly presented?	___	___
Well illustrated?	___	___

How do you use this publication in your job? Does it meet your needs? How can it be improved? Please be specific or cite examples:

_____

_____

_____

_____

_____

_____

Your name _____

Your title _____

Company name _____

Address _____

_____

FOLD THIS FORM, STAPLE, AND MAIL.
No postage is necessary if mailed in U.S.

Figure 1-1. Reader Comment Form

2
SECTIONS

The divisions that would be called *chapters* in a book are usually called *sections* in a technical publication. The sections of text form the body of the publication and include:

Page numbers
Running feet and heads
Headers
Text
Warning, cautions, and notes
Trademarks and logos
Footnotes

2.1 PAGE NUMBERS

Each section begins on a right-hand page. Pages are numbered in the lower outside corner by the section number, a hyphen, and the sequential number within the section, for example: **1-1**, **1-2**, **1-3**, **2-1**, . . . , **7-10**.

Why do you think numbering internally to each section would be more efficient than numbering the entire book sequentially? Consider first that your reader can find information more easily and second that you can update the page numbers more easily when changes are made only to some sections.

If your publications are printed on both sides of the paper, as is usual, then the page number must be flipped so that it always appears in the lower outside corner (right margin on odd-numbered pages; left margin on even-numbered pages).

NOTE: When sections are grouped by part titles, the pages naming the parts are not numbered. Each part is titled by an uppercase Roman numeral, a colon, two spaces, and the part title in caps, for example:

Part II: RANDOM LOGIC BLOCKS

A part may or may not contain additional text. If it does, then the page is numbered, and the running foot carries the title of the part, for example:

RANDOM LOGIC BLOCKS II-1

2.2   RUNNING FEET AND HEADS

Running feet are required to provide reference points to help your readers find information. Running heads should be used when there is a proprietary notice. Your company protocols might also require the company name and/or the date.

Running feet contain the section title on the same line as the page number, separated from the page number by two spaces. If the section title is too long to fit in the running foot, then consider shortening the title or using title keywords. Shortening is often the better choice.

The following two examples illustrate a running foot on an odd-numbered page and on an even-numbered page:

PHYSICAL COMPONENTS   3-1

3-2   PHYSICAL COMPONENTS

If sections are very long, then you might prefer to use subsection titles rather than section titles in the running feet.

2.3 HEADERS

Headers consist of (1) a number that places information at the correct sequential and emphatic level and (2) a title that describes the information to follow.

NOTE: Always follow every section header, no matter the level, by at least one line of text.

NOTE: Always repeat the information in the header in the first or second sentence of text. Don't implicitly refer to the header, for example:

1.2.3 Primary Menu
The Primary Menu lists three options: is correct, not [It lists three options] nor [This menu lists three options].

Each section may consist of up to four levels of headers in the following formats:

Level 1: n - SECTION TITLE

Level 2: $n.n$ SUBSECTION TITLE

Level 3: $n.n.n$ Sub-subsection Title

Level 4: $n.n.n.n$ Sub-sub-subsection Title

where n = number

NOTE: The word *section* may also be used generally to mean levels 1 through 4 when the distinction among levels is not important.

1.   Set level 1 (section) header words in caps centered about 1½ inches from the top of a right-hand page. Begin with the word SECTION, the section number, a space, a hyphen, a space, and the title, for example:

SECTION 3 - PHYSICAL COMPONENTS

Leave three blank lines before the first paragraph of text. Though the paragraphs to follow the level 1 header will usually introduce the topic of the section, don't include the title "Introduction" at the beginning of the first paragraph of a section (That is, don't use it as a level 2 header.) You may, of course, title a whole section (level 1) "Introduction."

2. Set level 2 (subsection) header words in caps. Begin at the margin with the section number, a period, the subsection number, two spaces, and the title, for example:

3.1   CENTRAL PROCESSING SYSTEM

Leave two blank lines before and one blank line after each level 2 header.

NOTE: If you use subsections, then include at least two per section.

3. Set level 3 (sub-subsection) header words in initial caps and underlines. Begin three spaces from the margin with the section number, a period, the subsection number, a period, the sub-subsection number, a space, and the title, for example:

3.1.1   <u>Instruction Unit</u>

Leave one blank line before and after each level 3 header.

NOTE: If you use sub-subsections, then include at least two per subsection.

4. Set level 4 (sub-sub-subsection) header words in initial caps. Begin five spaces from the margin with the section number, a period, the subsection number, a period, the sub-subsection number, a period, the sub-sub-subsection number, a space, and the title, for example:

1.1.1.1   Instruction Decoder

Leave one blank line before and after each level 4 header.

NOTE: If you use sub-sub-subsections, then include at least two per sub-subsection.

2.4 TEXT

The text under each header should be flush with the left margin for first and second level headers; indented three spaces for third level header; indented five spaces for fourth level headers. Unless your publications are double-spaced, do not indent at the beginning of paragraphs; just skip one line.

NOTE: A paragraph of technical text should not exceed ten lines.

2.5 WARNINGS, CAUTIONS, AND NOTES

Warnings, cautions, and notes allow you three degrees of emphasis. Warnings are at the high end, like first-degree murder and third-degree burns.

The office and art supply houses carry stickers and rub-ons for warnings and cautions. They generally stand out more forcefully than print.

2.5.1 Warnings

Warnings alert the reader to possible death or injury unless procedures are followed exactly.

The word WARNING and the warning appear, centered above the procedure, each time the procedure appears in the publication.

Safety precautions or their location in the publication should be stated, for example:

WARNING
The resonant capacitor contains more voltage than any other assembly in the drive. Follow all safety precautions listed in the first section of this manual.

2.5.2 Cautions

Cautions call the reader's attention to possible equipment damage unless procedures are followed exactly.

The word CAUTION and the caution appear, centered above the procedure, each time the procedure appears in the publication.

The reason damage might result should be stated, for example:

CAUTION
Before removing the air hose, vacuum the area around the filter discharge port to prevent contaminating the hose.

2.5.3 Notes

Notes notify the reader that a general rule is to be followed or that an out-of-the-ordinary condition exists. In general a note precedes the step it refers to in a procedure or follows a description or a list to note an exception or serve as a reminder.

A note begins with the word NOTE at the margin followed by a colon, two spaces, and the information, offset, for examples:

NOTE: **If the optional Port Frame is installed, then perform the following ten steps.**

NOTE: **If the system has returned a condition code other than "0," then print the outlist to find the error.**

2.6 TRADEMARKS AND LOGOS

A trademark designates a particular product brand. Company names, from Apple to Zilog, are considered trademarks also.

Use a trademark as an adjective, not a noun, for examples: **Super-10 CPU** and **Micromecca Super-10** are correct. [The Super-10's CPU] and [Micromecca's Super-10] are incorrect.

Don't distort the company logo. It is most easily recognized—which is its goal—when it always looks the same.

2.7 FOOTNOTES

Don't use footnotes in sections. First decide if the information is really necessary at all. If it is and will take no more than ten lines, then it can be included in a note. If the information is longer than ten lines, then include it in text or put it in an appendix.

3
NOTATION

Notation is any particular body of traditional written conventions assumed to achieve mutual understanding.

Technical writing requires particular precision and consistency in notation, including:

Punctuation
Acronyms and abbreviations
Numbers
Symbols
Computer communication
Spelling

3.1 PUNCTUATION

To produce the consistent punctuation necessary to technical text, be sure you understand basic grammar. If you are unclear as to the difference among simple, compound, and complex sentences; or between restrictive and nonrestrictive clauses; or between appositive and verbal phrases, see Chapter 11, "Grammar." Reviewing a composition text or handbook will help too.

True, many readers will be unaware of a lapse in punctuation, but others—people in hightech who read a lot—will be distracted at the drop of a comma. If your readers doubt your basic skills, then they may find it difficult to trust your explanations.

The punctuation mark usage rules listed in the following subsections are generally traditional, but some are specific to technical writing. The marks and rules are just a subset of the full set of each, but practicing the rules consistently will lend credibility to your publications. The following punctuation marks are discussed:

Period	.
Comma	,
Colon	:
Semicolon	;
Apostrophe	'
Hyphen	-
Dash	—
Slash	/
Parens	( )
Double quotes	" "
Underline	___
Ellipsis	. . .
Exclamation point	!

3.1.1 Period

Every declarative and imperative sentence ends in a period.

1. Use a period to end an imperative sentence, for example, this one.
2. Declarative sentences, for example this one, end with a period.
3. Use a period at the end of a punctuated list that forms a sentence, for example:

Use a period at the end of:
Imperative sentences,
Declarative sentences, and
Lists that form a sentence.

4. Don't use a period at the end of an abbreviation unless meaning would be ambiguous without it, for instance, *in* the preposition and *in.* the abbreviation for *inches.*

3.1.2 Comma

The comma is the most abused punctuation mark, maybe because it's the most used. Comma rules abound, but the following baker's dozen are mandatory.

1. Use a comma before a coordinating conjunction—*and, or, nor, for, so, but,* and *yet*—in a compound sentence when the independent clauses contain six words or more, so this sentence is an example of the rule.

   You may use commas in a series of short independent clauses, say of five words or less, for example: **Set the configuration switches, press the START/STOP button, and run the diagnostics.**

2. Separate each element in a parallel series of three or more items by a comma, for example: **the x, the y, and the z inputs** or **the x, y, and z inputs** not "the x, the y and the z inputs"—unless you want the "y and the z inputs" to be considered as a set separate from the "x."

3. Set off introductory dependent clauses by a comma, for example: **When the file is opened, the filename will appear at the top of the screen.**

4. Set off introductory phrases of three or more words by a comma, for example: **At the prompt, enter the coordinates by means of the mouse or keyboard.** This practice may have become less common in rhetorical text, but it is still a good idea in technical text.

5. Set off nonrestrictive clauses, appositives, and parenthetic expressions (not enclosed in parentheses) by commas when they are placed in the midst of a sentence and by a single comma when they are placed at the end, for examples:

   The START/STOP, which is the first button on the left, is lit when the system is running.

   The START/STOP button, the first button on the left, is lit when the system is running.

   Press the START/STOP button, the first button on the left.

6. Separate two or more coordinate adjectives that modify the same noun with commas when the adjectives could be logically joined by *and*, for example: **The sharp, colorful display makes graphs easy to design**.

   Sometimes the adjective immediately preceding the noun is so closely linked to the noun that it cannot be considered coordinate with the other adjectives, for example in **The high-performance IBM-compatible mainframe market is growing**, "high-performance" and "IBM-compatible" could not be joined by *and*.

7. Separate comparative structures with a comma, for example: **The faster the circuit, the more heat it generates**.

8. Set off contrasting words or phrases by a comma, for example: **The system is friendly, not clairvoyant**.

9. Use a comma to prevent misreading, for instance when words that often function as prepositions are used as adverbs, for example: **Beneath, the cables are connected to the floppy drives**.

10. Don't use commas around clauses that are restrictive.

11. Don't separate a subject from its verb by a single comma.

12. Don't introduce a series or end a series with a comma—unless the comma is required by some other rule; for example, in **The processor consists of the CPU, the memory, and the channels,** don't put a comma after "consists of."

   Did you notice that the last sentence ends a series with a comma? Do you see why? If not, then look at comma rule 4. And did you also notice that comma rule 12 and comma rule 10 exemplify themselves? Is that a useful (subliminal) technique that helps you to learn?

13. Use commas to mark off three-number units in numerics of four digits or more, for example: **2,851** and **18,371**.

3.1.3 Colon

A colon introduces a series, a list, an explanation, an example. In rhetorical text, a colon generally follows an independent clause. In technical text, for economy, a colon may also follow a dependent clause or a phrase that introduces a series or list, for example:

Parameters include: datapath widths, control signals, and cycle time.

3.1.4 Semicolon

A semicolon can substitute for a period, a comma, or a colon.

1. Use a semicolon before a conjunctive adverb that connects independent clauses and a comma after, for example: **The RAM has 64 inputs; however, they do not all carry data**.

2. Use a semicolon between independent clauses when the clauses are logically and grammatically parallel and when no coordinating conjunction is used, for example: **The SAVE command writes the data to disk; the QUIT command does not.**

3. Use semicolons in a sentence that contains a series of clauses or phrases having internal commas, for example: **EXIT commands include SAVE, which writes the data to disk; QUIT, which does not; and QQUIT, which gives you another chance.**

4. Use semicolons in a sentence that contains a series of numbers having internal commas, for example: **1,582; 603; 170,040; and 17,381**.

3.1.5 Apostrophe

An apostrophe shows possession or contraction, *never* pluralization. You probably won't be using contractions much unless your company standards allow a casual second-person tone.

By the content and the punctuation of the last sentence, how would you characterize the tone of this book? What does that say about how I view my audience—you?

1. Use an apostrophe and an *s* to indicate possession in a singular noun, for example: **The instruction's second operand.**
2. Use an apostrophe to indicate possession in a plural noun that ends in an *s*, for example: **several instructions' operands**.
3. Use an apostrophe and an *s* to indicate possession in a number (usually a year or a decade) used as a noun, for example: **1970's models were primitive**.

    Generally, it is better to recast the sentence so that a number won't need to show possession, for example: **The models of the 1970s were primitive**.

    NOTE: Numerical years and decades do not take apostrophes to show pluralization.

4. Do not use an apostrophe for the possesive pronoun *its*. It's the same as *his* or *hers*.
5. Don't use an apostrophe for pluralization of acronyms, for example: **Two CPUs can be lashed together** is correct.

    NOTE: The plural *s* is lower case.

    Did you notice the inconsistency between "Do not" and "Don't" in the two rules above? Now that you can assume I did it on purpose, does the "do not" carry enough more weight than "don't" to stress the absoluteness of the rule? If not, then this consideration might.

3.1.6 Hyphen

A hyphen joins two or more words that act as a noun or adjective or makes a word easier to read.

1. Hyphenate to form a compound adjective when the first word modifies the second, for examples: **low-intensity** and **high-performance**.
2. Hyphenate multiple-word terms that immediately precede a noun, for example: **two-to-one ratio**.
3. Hyphenate to prevent misreading, for instance between "re-creation" and "recreation."
4. Hyphenate a duplicate prefix, for example: **sub-subassembly**.
5. Hyphenate a number-unit adjective, especially those involving bits, bytes, or pins, for examples: **64-bit bus, 16K-byte block,** and **10-pin package.**
6. Hyphenate compound nouns naming a unit of measurement, for example: **foot-pound** and **kilowatt-hour**.
7. Hyphenate to prevent doubling a vowel or tripling a consonant, for examples: **re-entry, pre-existing, co-operation,** and **brass-smithing.** If you prefer to follow Webster, then you won't hyphenate the first three, but whatever way you decide, be consistent.
8. Hyphenate a mixed number between the integer and fraction, for example: **3-1/3**.

3.1.7 Dash

Dashes come in two flavors: m and n. The m-dash is the size of a printed m; the n-dash is half as long. You will probably use two hyphens for the m and one for the n.

1. Use m-dashes to set off a series embedded in a sentence, for example: **Specify the output--print, punch, or display--and press ENTER.**

2. Use m-dashes to set off a parenthetic expression when you want to convey emphasis, for example: **Connect P-23--the power plug--to J-23.**

3. If you use an m-dash rather than a colon to introduce a list, then use an m-dash to introduce every list in the book.

4. Don't use an n-dash as a substitute to mean *to* or *through* in sequential numerics in text; spell out the words because the n-dash might be read to mean either. For example, **instruction lines 19 to 30** means not including 30; **instruction lines 19 through 30** means including 30; but "instruction lines 19−30" could mean either, unless there are only 30 lines.

3.1.8 Slash

Slashes are seldom seen in rhetorical prose, but that is no reason to slander them in technical prose. They are economical and, as long as they are used consistently, clear. *I/O*, slash included, has become a standard acronym.

1. Use a slash to mean *per* in a table, for example: **ops/sec**; but use the word *per* in text, for example: **two minor cycles per cycle.**

2. Use a slash between two acronyms, for example: **CAD/ CAM.**

3. Use a slash between *and* and *or* to mean *both or either*, for example: **Write the file to tape and/or to the screen.**

4. Use a slash between the two words of a compound term to mean *and*, for example: **programmer/analyst**.

5. Use a slash between two words to mean *or* only when the relationship is an obvious opposition; otherwise use *or*, for examples: **ON/OFF** and **print or display**.

3.1.9 Parens

1. Parentheses, parens, enclose an acronym the first time that it's used (following the spelled-out term).

2. Use parens sparingly, especially in software manuals, because they are often used as part of a command.

3.1.10 Double Quotes

Double quotes enclose titles of portions of larger works, and they may also be used around specific terms in text.

1. Use quotes around section, figure, or table titles, but not around references to their numbers alone, for example: **See Section 1.3, "Cabling," for hook-up procedure**, not [See "1.3"] nor ["Section 1.3"].
2. When you refer to commands or computer messages in text, don't put periods and commas inside quotes as you would in rhetorical text, for examples: **Type "Logon", and the Super-10 will respond with "Your ID please:" and Enter "ABC.703", the type coordinates, and the revision level, and press ENTER.**

3.1.11 Underline

Underlines stand in for italics. If your system does not have the capability for italics, then:

1. Underline all words in the title of a referenced publication, whether technical, trade, or text book.
2. Underline uncommonly used foreign words like *in vitro*, but don't underline those commonly used like "vice versa" and "versus."

3.1.12 Ellipsis

The three dots that form an ellipsis may be placed horizontally or vertically.

1. Use ellipses in command, operation, and instruction se-

quences to indicate that material has been left out, for example:

Login.
Enter Password.

.
.
.

Request session history.
Logout.

2. Use ellipses in commands and operations to show that the objects or actions are iterative, for example: **Create the directory by entering "dirname, filename, . . . , filename"**

3.1.13   Exclamation Point

Avoid using exclamation points even in marketing publications.

3.2   ACRONYMS AND ABBREVIATIONS

Acronyms and abbreviations abound in technical texts. Don't fight it because they are the most economical way to name the complex set of concepts, operations, and measurements that form the technical dictionary. They are more quickly read and said.

Some have reached the status of words: **bit**, for example, is the shortened form of **b**inary dig**it** (or **bi**nary dig**it**), and who says **kilo** when **K** will do?

In fact, acronyms and abbreviations are so prevalent in the electronics industry that people often don't know the original spelled-out terms the acronyms stand for. It's so much easier to read or talk about "MOS/VLSI" than "metal-oxide-semiconductor very large-scale integration."

3.2.1   Acronyms

An acronym is generally formed from the initial letters of the words in a compound term.

Strictly speaking an acronym can be pronounced as a word, for examples: **RAM** for **r**andom **a**ccess **m**emory and **DOS** for **d**isk **o**perating **s**ystem.

Strictly speaking a group of letters pronounced like initials is called an initialism, for examples: **CPU** and **I/O**.

However, the distinction is fading, and some acronyms may be pronounced either as a word or initials, for example **ALU** (arithmetic logic unit) and **EMA** (electromagnetic actuator).

To blur the distinction still further, shortened forms of computer instruction opcode names, called mnemonics, also look like acronyms and initialisms, for example: **SAC** for Set Address Space Control and **MVN** for Move Numerics.

CPU is one of the few of the industry standard acronyms; it hardly needs to be spelled out anymore, ditto for *I/O* and *RAM*; but spelling them out at first use is also standard.

Generally acronyms are assigned to software products and software and hardware functional components. When you use an acronym:

1. Capitalize all letters in the acronym whether or not the words in the compound term are capitalized.

2. Put the acronym (in parens) immediately after the first use of the spelled-out term. If sections of the publication are apt to be read by different audiences, then repeat the term and acronym at first use in each section.

3. Use the acronym consistently after it has been defined instead of switching back and forth between the acronym and the spelled-out term.

4. The choice between indefinite articles depends on the spoken word. Use *a* before acronyms beginning with a consonant sound and *an* before acronyms beginning with a vowel sound, for examples: **a CMOS device** and **an NMOS device**.

5. Only possessive acronyms take an apostrophe before the *s*; there is no apostrophe between an acronym and the plural *s*, for example:

    CPUs = two or more
    CPU's = belonging to one CPU
    CPUs' = belonging to two or more CPUs

Where else would you expect to find this rule, under "Apostrophe"?

6. If more than one term in the publication reduces to the same acronym, then change one term, or use the acronym for only one of the terms.

7. Put all acronyms and their spelled-out definitions in the glossary.

8. Don't assign an acronym to a term that you will use only two or three times in the publication.

3.2.2 Abbreviations

An abbreviation is the shortened form of a word, composed of letters selected from the full word. Most abbreviations are not spoken, but some like *K* and *ac* are. When you use an abbreviation:

1. If the first letter of the word is capitalized, then capitalize the first letter of the abbreviation.

2. Use the same abbreviation for a measure whether the amount is singular or plural, for example: **1 cm to 12 cm**.

3. Spell out *that is*, not [i.e.], and *for example*, not [e.g.].

4. Spell out *versus*, not [vs.].

5. Spell out *maximum* and *minimum* in text; abbreviate them to *max* and *min* only in tables.

6. Don't use a period at the end of an abbreviation in a table.

7. Don't use a period at the end of an abbreviation in text unless the meaning would be ambiguous without it, for example, the preposition *in* and the measurement *inch*. Better yet, spell the term out.

8. Don't use an apostrophe or a period for clipped forms, like *phone* (not ['phone]) and *lab* (not lab.). They are becoming words on their own.

The IEEE publishes a comprehensive list of approved abbreviations. Table 3-1 lists some of the more common.

Table 3-1. Common Abbreviations

ac	alternating current
A	ambient
A	ampere
A/D	analog to digital
Btu	British thermal unit
c	centi- (hundredth)
cu	cubic
d	deci- (tenth)
D/A	digital to analog
da	deka- (ten)
dB	decibel
dc	direct current
G	giga- (billion)
gnd	ground
h	hecto- (hundred)
Hz	Hertz
ips[a]	inches per second
k	kilo- (thousand)
K	kilo (1,024)
M	mega- (million)
m	meter
m	milli- (thousandth)
n	nano- (billionth)
p	pico- (trillionth)
psi	pounds per square inch
s	second
sq	square
V	volt
W	watt

[a]IPS is the acronym for instructions per second.

Many of the abbreviations in the table can be mixed and matched. The prefixes in the table—*c, d, da, G, h, k, M, m, n,* and *p*—can be combined with the units of measure to make smaller or larger units. For examples: **ps** is picosecond, a trillionth of a second; **kW** is kilowatt, a thousand watts.

3.3 NUMBERS

Numbers can be spelled out in alpha characters or expressed in numeric characters. Be especially careful in proofreading numerics. When you use a number:

1. Spell out ordinal numbers like *first, second, third*.

2. Spell out cardinal numbers *one* through *ten*, and use numerics for *11* and above in paragraphs of text; however, when numbers one through ten and above are mixed in one sentence, express all the numbers in one form, for example: **4 boards, 16 connectors, and 48 DIPs**.

3. Spell out numbers at the beginning of a sentence. If the number is awkwardly long or if there are other numbers in the sentence, then rewrite so that the number does not come first. Maybe it can follow a semicolon.

4. Use a hyphen in spelled-out compound numbers, for example: **twenty-two**.

5. Spell out approximate amounts, for example: **hundreds of instructions**.

6. Spell out *million, billion*, and *trillion*, but express *1,000* in numerics—unless it's approximate.

7. Express bit values in numerics *1* and *0*.

8. Express percentages in numerics with a percent sign, for example: **73%**. Spelling out *percent* just takes longer to read.

9. Join mixed numbers (integer-fraction) by a hyphen, for example: **3-⅓**.

   Where else would you expect to find this rule, under "Hyphen"?

10. When two numbers in succession modify one noun, spell out the first number and express the second in numerics, for examples: **sixty-four 8-byte words** and **twelve 100-pin connectors**.

11. Use commas to mark off each three-number unit in numerics of four digits or more, and use semicolons to separate a sequence of such numbers, for examples: **1,582; 603; 170,040; and 17,381**.

    Where else would you expect to find these rules, under "Comma" or "Semicolon"?

12. If a decimal point value has no preceding integer, then use a preceding zero, for example: **0.493**.

13. Use either × or *by* to link dimensions, but use your choice consistently, for example: **2 cm × 3 cm** or **2 cm by 3 cm**.

14. If your publications include both metric and British measurements, then use one first and the other (in parens) following consistently, for examples: **1 in (2.54 cm)**, **1 sq in (6.452 sq cm)**, and **1 square foot (0.092 square meters)**. If you abbreviate one of the measurements, then abbreviate the other.

15. Don't express the same number twice as a word followed by a numeric; for example, [ten (10)] is overkill.

3.4 SYMBOLS

A symbol is a conventional sign that represents a specific component, operation, or relationship in schematics, logic diagrams, flowcharts, and algorithms.

You will be using symbols in most publications, for instance, ground symbols in installation manuals, fuse and test point symbols in maintenance manuals, latch and memory circuit symbols in troubleshooting manuals, flowchart symbols and algorithms in software manuals.

Symbols vary from company to company and from designer to designer, but they must be reconciled for company publications so that you can present them consistently.

A template serves to keep flowchart symbols consistent. (A subset of flowchart symbols is included in Section 4.5, "Flowcharts.") The IEEE standard is best for operator symbols. A subset of operator symbols follows in Table 3-2.

3.5 COMPUTER COMMUNICATION

Ergonomics, the study of person and machine interaction from the person's point of view, is intended to help people be comfortable working with electronic machines. The more a person can learn about communicating with the machine, the easier the task.

Table 3-2. Operator Symbols

Symbol	Meaning
≈	approximately
:	compared to
÷ or /	divided by
=	equal to
⩾	equal to or greater than
⩽	equal to or less than
>	greater than
<	less than
−	minus
× or *	multiplied by
¬	not
≠	not equal to
Ω	ohm
→	yields, forms

In technical publications, the clear and consistent presentation of computer instructions and commands helps readers to learn and remember them.

Strictly speaking instructions and commands are different in that *instruction* means a coded directive, and *command* means an electronic signal; however, both are entered electronically now so that distinction has faded. Now, *instruction* usually refers to assembly language and to application languages that are noninteractive, for instance, FORTRAN, PASCAL and PL/1, while *command* refers to computer languages that are interactive, for instance, query systems, graphics systems, and editors.

Instructions need generally be discussed only in programming manuals, but commands are included in almost all operator's manuals, and installation and troubleshooting manuals often need commands and codes for bring-up and diagnostics. Both instructions and commands are similar to numbers in that they need especially careful proofreading.

NOTE: In the following discussions, *commands* is used as a general term including both *commands* and *instructions*.

3.5.1 Command Notation

Like all languages, computer languages are governed by syntax and diction rules that structure their meanings. In addition, you will superimpose command notation, which consists of both the language and the symbols you choose to show whether elements are optional, required, constants, variables, defaults.

The notation conventions need to be included so that the reader can understand the requirements of that command.

Table 3-3 is a sample of notation for commands.

Use the same notation throughout all publications so that your readers can become familiar with it, and include a command notation table using a super or subset of the same notation in each publication.

3.5.2 Command Format

The command format is the way in which you describe a command visually. As with notation, use the same format throughout the publication. However, the format will probably be different for external (interactive) programs and internal (system) programs.

The command name, definition, and syntax are necessary to all command descriptions. The other categories depend on the commands you are describing. The following two sections contain samples of format for program externals and program internals.

3.5.2.1 Program Externals

End-users are the most frequent readers of program externals as they first learn to use the system and then expand their capabilities to use more of the system's capabilities.

See the top of page 43 for an example of a text editing command description. A few optional categories are included in the example.

Table 3-3.   Command Notation

lowercase	Characters to be replaced with a user-selected variable. For example, **mmddyy** (month day and year, two characters each) might be entered as **052586**.
UPPERCASE	Characters that must be entered as shown. For example, **Browse** is a command whose first character must be entered, but the rest of the command need not.
< >	An element to be entered at the keyboard.
/ , . ( )	Punctuation elements that must be entered as shown. For example, <**userid.filename.filetype**> means that variables that replace the elements "userid," "filename," and "filetype" must be delimited by periods.
|	Separators between members of a group of choices. For example, <**Y|N**> means there's a choice between the two.
[ ] $\begin{bmatrix} \ \end{bmatrix}$	Enclose a group of choices, one of which is required. For example, **[10K| 20K| 48K]** means that one of the three must be chosen.
{ } $\left\{ \ \right\}$	Enclose a group of choices, none of which is required. For example, {**size | view**} means that one or both options may be entered. $\left\{ \begin{matrix} \textbf{tape} \\ \textbf{view} \end{matrix} \right\}$ means that one option may be entered, but none need be.
_____	Indicates a default in a group of choices. For example, **ON** | **OFF** means that the function is on unless "OFF" is entered.
. . .	Indicates elements that may be repeated. For example, <**DELete filename, . . . , filename**> means that several variables may be entered, that variables must be separated by commas, and that "DEL" is the minimum required characters of the command.
" "	Indicates a command or term mentioned in text. For example, **The following sections describe "Edit" options.**

EDIT Command

Definition:	Accesses filespace to create, copy, or change text.
Syntax:	<Edit filename.filetype.extent>
where:	filename - the name of the file you want to create or modify, 8 alpha characters.
	filetype - "T" for text or "P" for program.
	extent - 3 alphanumeric characters.
Keys:	"F3" to save and "F4" to file.
Restrictions:	The system must be in EDIT mode.
Comments:	The logon ID automatically controls all files under that ID. No one else can modify the file.

Does the command notation within the format match the notation in Table 3-3?

3.5.2.2 Program Internals

Systems programmers and application programmers are the most frequent readers of program internals as they learn to code in new languages and to control system capabilities.

Systems programmers still seem to like uppercase characters. It's the mystique. The following is an example of system initialization parameter description.

SAVESYMACRO	
DEFINITION:	SPECIFIES PHYSICAL AND LOGICAL PARAMETERS NEEDED TO SAVE SYSTEM ON DISK.
SYNTAX:	SAVENAME
	SYSZE=NNNN
	SYSTRT=(CC, P)
	SYPGNM=(NN, . . . NN, NN-NN, . . .)
	PROTECT = $\left\{ \begin{array}{l} ON \\ \overline{OFF} \end{array} \right\}$
WHERE:	
SAVENAME	- THE NAME UNDER WHICH SYSTEM IS TO BE SAVED.

SYSZE - THE MINIMUM STORAGE SIZE NEEDED TO SAVE THE SYSTEM.

SYSTRT - THE STARTING CYLINDER (CC) AND PAGE ADDRESS (P) ON WHICH THIS NAMED SYSTEM IS TO BE SAVED. SPECIFY IN DECIMAL.

SYPGNM - THE NUMBERS OF THE PAGES TO BE SAVED (NN). SPECIFICATION MAY BE DONE AS SINGLE PAGES OR AS GROUPS OF PAGES. FOR EXAMPLE, IF PAGES 0 AND 7 AND 10 THROUGH 13 ARE TO BE SAVED, THEN USE THE FORMAT: SYSPGNM = (0, 7, 10-13).

PROTECT - TO RUN EITHER PROTECTED (ON) OR UNPROTECTED (OFF). ON IS THE DEFAULT. IF A NAMED SYSTEM IS SPECIFIED AS UNPROTECTED, THEN CHANGES WILL NOT BE DETECTED BY THE CONTROL PROGRAM, BUT WILL BE SEEN BY ALL USERS OF SHARED DATA.

ATTRIBUTES: PAGEABLE
ENTRY POINT: SYINITPARM

EXAMPLE: SAVSY1
SYSZE=4096
SYSTRT=(40, 1)
SYSPGNM=(0, 7, 10−13)

3.5.3 Error Messages

Error messages should also be presented consistently. The message should be printed exactly as it is shown, whether on the terminal screen or in the program listing; then the possible conditions, the meaning of the error; then the user actions in response; then if the system responds to the user action, include the responses, for example:

Message:	--- DISK SPACE FULL ---
Conditions:	The file exceeds available storage space on the floppy disk, or Drive 1 door is not closed.
Action:	Check that Drive 1 door is closed. If Drive 1 door is closed, then load a new floppy disk into Drive 1, and "Save" the file.
Response:	FILE filename IS SAVED

Notation of commands that the user performs in reaction to system messages should also be consistent with the notation used for commands in application and operating system programs.

3.5.4 Pseudocode

Pseudocode is a clever notation invented much like a pidgin English, part programming language and part English.

Pseudocode is a helpful notation for describing program logic. The following logic, for example, causes path selection after a branch:

```
IF TBL hit,
  THEN BEGIN
    Initiate predicted target fetch;
    AGEN target address
    IF target address NOT = predicted address
      THEN BEGIN
        Fetch actual target instruction stream;
        Cancel predicted stream
      END;
  END;
ELSE
  IF I-Unit predict taken,
    THEN initiate target fetch
  ELSE initiate nonsequential fetch;
    IF branch falsely predicted taken,
      THEN BEGIN
        Cancel predicted stream
        Switch to sequential stream
      END:
END.
```

3.5.5 Summary Card

Summary cards are helpful for programmers and operators who need a handy reference to command and instruction languages.

An 8½ by 11 inch sheet of light cardboard accordion-folded into six panels is the most common sized summary card. The front panel should hold the name of the system, a short paragraph about its function, a reference to the publication that the card accompanies, and the command notation. The other five panels list the syntax of as many commands as will fit legibly on each. Of course larger systems require larger cards.

Figure 3-1 illustrates a summary card.

3.6 SPELLING

Even with all the rules for punctuation and abbreviation, the standards for alpha, numeric, and symbolic representation, and the forms for computer communication, you will still have some choices to make in spelling.

First, keep a dictionary handy. Then keep a technical dictionary handy, one that covers electronics terms used in your part of the industry. If you are writing for a company that makes a product compatible with another company's product, then keep that company's glossary handy, too, and use the same spelling for the same terms.

In any case, decide—department wide—on the spelling for words that might be abbreviated or capitalized, that could be spelled as one word or hyphenated, that have more than one correct spelling. Decide on acronyms and abbreviations that the department will and won't use. Make a list and keep it updated.

Every writer should have a copy of the list in order that notation be consistent throughout all text in all publications.

The following is a sample standard spelling list.

air-cooled	back-up (as noun or adjective)
analyze	bi-directional
appendixes	bottom-up
back up (as verb)	bring-up

<EDit MODel
<EDit Signal

MICROMECCA

EDITOR

COMMAND SUMMARY

This card provides a quick reference summary
of Editor (ED) commands. Commands
are described in the Editor User's Guide,
002467-001.

*   An asterisk means "all." Example:
     QUEry SYmbol * userid   means list
     all symbols under the named userid.

| Vertical bars indicate multiple
    choices. Example: OFF | ON

UPPERCASE   Uppercase letters indicate an
            element that must be typed as
            shown or an abbreviation.

lowercase   Lowercase letters indicate a
            variable. Example: mmddyy
            could be typed as 010584
            (January 5, 1984).

{ }   Braces indicate required
      elements. Do not type.

[ ]   Brackets indicate optional
      elements. Do not type.

_____   Underlines indicates default
         values. Example: OFF | ON.

Figure 3-1.   Summary Card

cannot
check-stop
crosshairs
database
data set
disk
doubleword
error-free
fan-in
fan-out
fetch
firmware
fixed-point
flip-flop
floating-point
fullword
Gbytes
halfword
hardcopy
high-density (as adjective)
high-performance(as adjective)
high-speed (as adjective)
ID
indexes
Kbytes
logoff (or logout, dependent
  on the system
logon (or login, dependent on
  the system)
maximum
microcode
microelectronics
microprogram
microstore

microword
minimum
minutes
multi-bit
multiplex (as verb)
multiplexer (as noun)
multiprocessor
mux (verb; singular = muxes)
MUX(noun; plural = MUXes)
nonzero
opcode
power-on
pseudocode
read
runtime
setup
single-cycle
state-machine
store
subatomic
subroutine
top-down
usable
versus
workspace
workstation
write
nnn-bit (for example, 16-bit
  word)
nnn-byte (for example, 64-byte
  line)
x-compatible (for example,
  DEC-compatible)

If you have a spelling checker in your word processing system,
then add the spelling words to its dictionary.

4

FIGURES

A figure is a graphic representation of information such as an illustration, a chart, a diagram, or a photograph.

Like the text, figures need to be directed to your audience. Do your readers need to understand the overall structure of a system? A hierarchy chart or block diagram will show it. Do they need to know how to assemble or disassemble parts of a machine? An exploded illustration is a good reference. Do they need to follow a complex procedure or the logic in a program? Flowcharts are in order.

Your choice of figures is based not only on the topic, but also on what equipment is available and what has been created by product developers. You will probably modify and simplify schematics and mechanical drawings generated by designers more often than you will create the figures from scratch.

Regardless of their content, figures benefit from the following considerations:

1. Don't expect any figure to be totally self-explanatory.

    Introduce a figure by its content, for example: **Figure 1-3 illustrates the CPU components**. Figure 1-3 should then be titled "1-3. CPU Components," and should illustrate all and only the CPU components that are mentioned in the text.

2. Label consistently.

   Use callouts identical to terms used in text. If a figure needs symbols, acronyms, or abbreviations other than those used in text, then explain them in a key on the figure.

3. Keep in mind that the publication may undergo revision.

   Design the figures with as much generality as possible, leaving out the details, which are most likely to change.

4. Use typographic techniques.

   Depending on the equipment available, design figures to take advantage of contrasting line weights and character fonts.

   For example, use the boldest, largest lettering and line weight for the most important points in a figure, and decrease the font size and line weight according to descending order of importance.

5. Size the image.

   Leave enough white space around the figure to separate it from text.

   Consider the size of print in the figure; it may not be legible after reduction to book or microfiche size.

6. Center a figure title under each figure.

   Set the word *Figure* and each word of the figure caption in initial caps. Number the figure with the section number, a hyphen, the figure sequence number within a section, a period, and two spaces. For example:

   Figure 2-1.   Execution Unit

7. If figures are members of a set, make their relationship clear.

Center *n of n* (in parens) under the figure title to link them. For example:

Figure 1-1.   System Flow
(2 of 3)

8. Don't design figures that make the readers turn the publication sideways.

   Almost all figures can be set up to fit your page size or to fill multiple pages.

The following sections discuss figures under the categories:

Hardware illustrations
Screens
Hierarchies
Block diagrams
Flowcharts
Photographs

4.1   HARDWARE ILLUSTRATIONS

Text dealing with hardware assembly or maintenance benefits greatly from illustration, which is to a large extent language independent.

Hardware drawings may be either isometric or orthographic views. Isometric, three-dimensional, views are best used to stress spatial orientation, including cabinet assemblies and machine adjustment directions.

Orthographic, two-dimensional, views are best used to stress details, including equipment floor-plans and machine adjustment details.

Orient the viewer. When the subject of the figure is a part of a larger structure, show the relationship either by a sequence of figures or by a blowout (a telescoped enlargement of the part under consideration).

Figure 4-1 illustrates a blowout; Figure 4-2 illustrates the subject of the blowout; and Figure 4-3 illustrates detail of the subject.

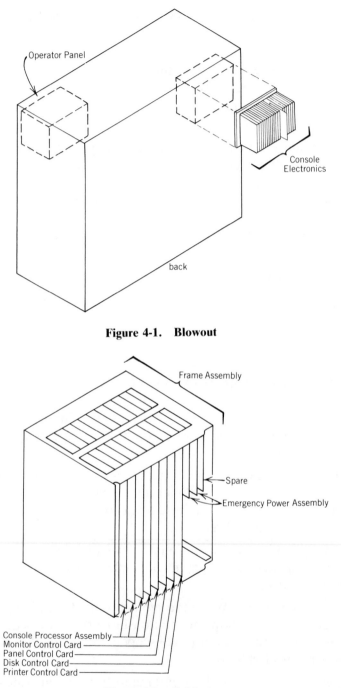

Figure 4-1. Blowout

Figure 4-2. Subject

4.2 SCREENS

End-users like to be able to check their operations. They are reassured to read, for example: **The following screen will be displayed** and then to see that the screen illustration in the publication matches the display on the screen they are looking at.

Screen illustrations should be presented exactly as the display appears and should be boxed off from the text.

If there are many screens in one book, for example, a user's guide for a data processing system, it is not necessary to designate each screen by a figure number and title—especially if you don't have automatic numbering—but introduce the display consistently, for example: **The following display allows you to . . .**

Figure 4-4 illustrates a screen display.

If any of the fields on the screen are unexplained in the display itself, then explain them following the figure, for example:

where:
EFFECTIVITY—"N" for new or "O" for old. Default is "N".

4.3 HIERARCHIES

Hierarchy charts are useful only to illustrate natural hierarchical structures, for example, machine assemblies, hardware or software functional components, nested subroutines, hierarchical databases, and departments in an organization.

Figure 4-5 illustrates an assembly hierarchy, sometimes called a "family tree." Family trees are generally created by draftspeople to show all the parts of a system that bear a part number. You can choose the parts of the tree that you are writing about.

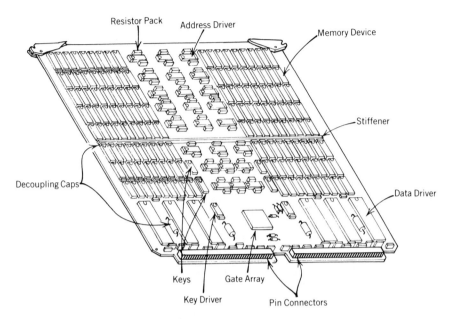

Figure 4-3. Detail

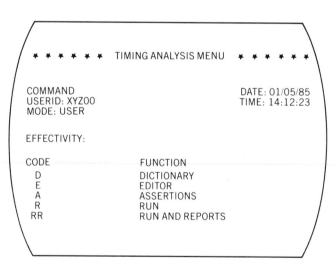

Figure 4-4. Screen Display

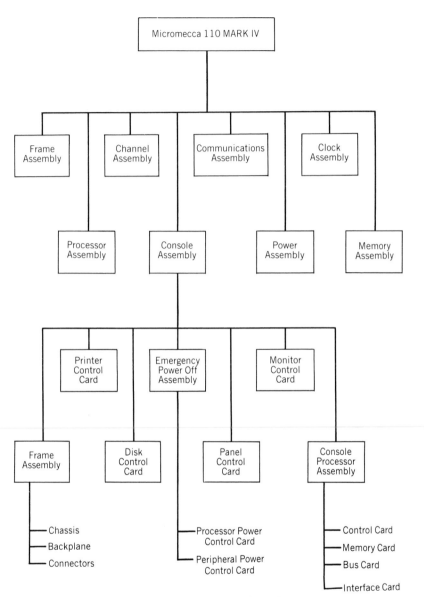

Figure 4-5. Assembly Hierarchy

4.4 BLOCK DIAGRAMS

Block diagrams best illustrate a group of black boxes and the relationships between them. Block diagrams stress parallelism and data flow rather than hierarchy; however, each black box is usually a part or a parent of another black box, much like a Russian doll, so a hierarchy chart is often the best introduction to a set of block diagrams.

Figure 4-6 is a high-level block diagram of functional units in a computer system.

4.5 FLOWCHARTS

Flowcharts best illustrate alternatives or a system with a large number of inputs and outputs. Flowchart symbols should be standard throughout all your company's publications. As in many other DP areas, IBM has pretty much established the standard. Table 4-1 lists and illustrates the most common flowchart symbols.

Flowchart symbols may be used to illustrate program execution, processes, and hardware configurations.

Logic flowcharts show the location of decision points and their possible outcomes (yes/no, pass/fail, on/off, I/O, match/no match) in the flow of program execution.

Processes flowcharts show the sequence of procedures and operations in a system that may contain hardware, software, people, paper, or all four.

The flowchart symbols denoting physical entities—tapes, disks, documents, signals—are used in figures illustrating hardware configurations.

Figure 4-7 illustrates the program flow in a port switch-handler routine; Figure 4-8, process flow in design and manufacture of a chip from design to ship; and Figure 4-9, a data processing system hardware configuration.

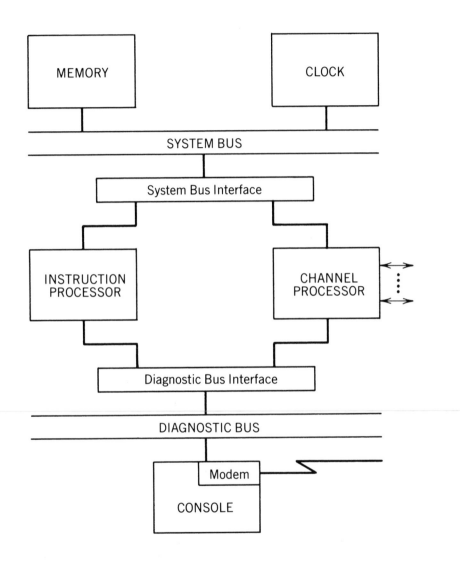

Figure 4-6. System Block Diagram

Table 4-1. Flowchart Symbols

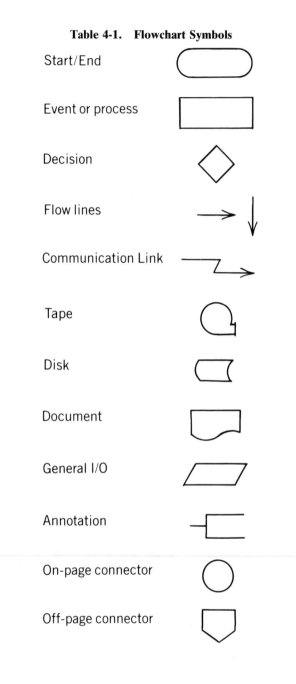

Start/End

Event or process

Decision

Flow lines

Communication Link

Tape

Disk

Document

General I/O

Annotation

On-page connector

Off-page connector

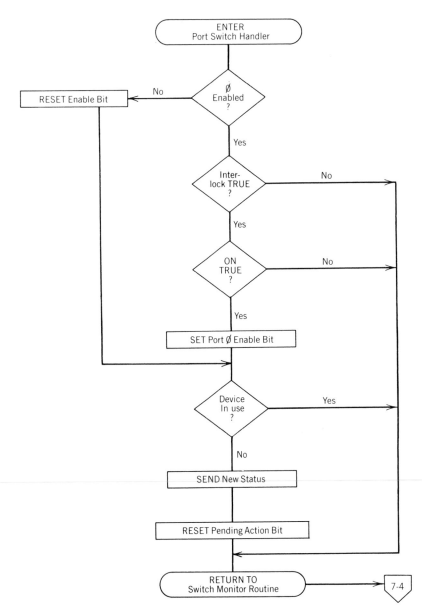

Figure 4-7.   Program Flow

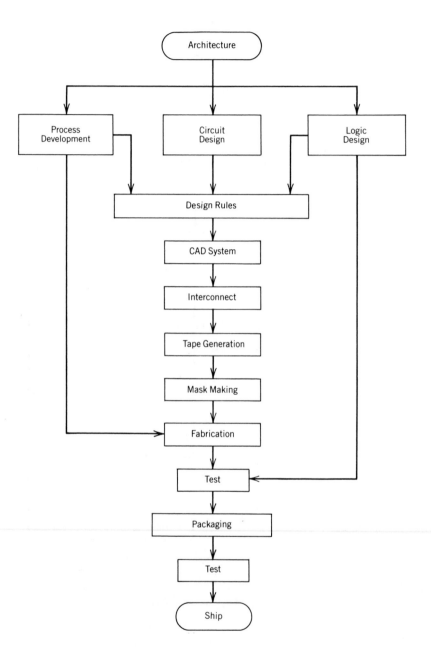

Figure 4-8.   Process Flow

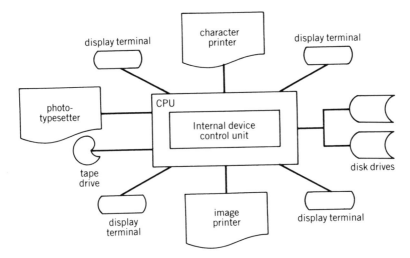

Figure 4-9. System Configuration

4.6 PHOTOGRAPHS

Photographs are often included in marketing publications and sometimes in hardware maintenance manuals.

When the product is visually exciting, a photograph lends a nice touch as a frontispiece in marketing publications.

Photographs can take the place of isometric drawings in hardware publications, and may be further clarified with callouts for the parts in question.

5

TABLES

A table is a representation of information in multi-column, often numerical, form. The rows and columns provide a system of classification; the column head identifies the items in the columns, and the first column, "the stub," identifies items in the rows.

To some degree, table format varies dependent on the information, but in general, the following considerations apply:

1. Center column heads in initial caps over each column.
2. Align words in columns by the first character.
3. Align numbers in columns by the decimal point or comma.
4. Footnotes can be helpful in a table because of limited space. In software publications, indicate footnotes by a symbol other than an asterisk because asterisks are commonly used in program statements. In fact, it may be difficult to find a character on the keyboard that isn't used in some notation or other.
5. Center a table title above each table.

   Set the word *Table* and the title in initial caps. Number the table with the section number, a hyphen, the table sequence number within the section, a period, and two spaces, for example: **Table 2-3.   Opcodes**.

6. If tables are continued over more than one page, then repeat the table title and label all repetitions with "CONT'D." (in parens) to link them, for example: **Table 3-2.   SYSGEN Parameters (CONT'D.)**.

7. If tables are fairly simple and there are many of the same kind in one book, then format the tables consistently, but it is not necessary to designate each table in a large set by a table number, especially if you don't have automatic numbering. The following truth table for a 2-to-1 MUX with enable is an example of a form that could be used as boilerplate for a book on circuits:

2-to-1 MUX						
S1	S0	A	B	F1	F2	Notes
0	0	X	X	0	1	disabled
0	1	0	X	0	1	
0	1	1	X	1	0	
1	0	X	0	0	1	
1	0	X	1	1	0	
where X = don't care						

8. Use space to separate columns and entries generously.

9. Use horizontal and vertical lines to separate columns and entries sparingly; the lines can clutter as well as clarify.

10. Don't design tables that make the reader turn the publication sideways.

Many kinds of information can be presented in a table: system requirements, parameters, part names and numbers, pin definitions—the list is endless.

Table 5-1 lists the specifications for a robot system.

Table 5-2 is a summary of the format and spacing standards for headers, discussed in Section 2.3.

Compare the table with the text. What information is lacking from the table? Would you add it? How?

Table 5-1.   Specifications

Element	Description
System	Five-axis robot gripper arm with control interface and power supply
Controller	Micromecca-M1 point-to-point control with segmented path control
Memory	4K EEPROM 4K RAM 232-point ROM
Drive	Open loop with automatic homing Electronic stepper motors
Interface	Dual RS-232
Power requirements	110 VAC or 220 VAC
Operating environment	110°F (38°C) limit
Cables	Robot control = 10 ft (3 m) Operator control = 10 ft (3 m)
Reach	18 in. (457 mm)
Speed	0−20 ips (0−508 mm/s)
Positioning accuracy	±0.020 in. (0.50 mm)
Gripper	Force programmable 1 lb−20 lb

Table 5-2.   Header Standards[a]

Format[b]	Spacing
SECTION n - TITLE	Center 1½ inch from top of page; followed by 3 blank lines
At least one line of text	
n.1 TITLE	Begins at margin; preceded by 2 blank lines, followed by 1 blank line
At least one line of text	
n.1.1 Title	Begins 3 spaces from margin; preceded and followed by 1 blank line
At least one line of text	
n.1.1.1 Title	Begins 5 spaces from margin; preceded and followed by 1 blank line
At least one line of text.	

[a]No level may have less than two members.
[b]Where n = number.

6
LISTS

Lists are simply information presented in vertical format to introduce or enumerate a sequence of topics.

Lists may be constructed of words, phrases, clauses, or sentences; they may contain no notation, or their entries may be joined by punctuation or prefixed by numbers, hyphens, or bullets.

The following numbered list discusses the considerations due lists:

1. When you introduce information with a list, be sure that the topics are listed in the same order that they are discussed in the text—and vice versa.

2. Choose one grammatical structure—words, phrases, clauses, or sentences—and use it consistently for each entry in the list.

3. Use all caps or initial caps in a list of words.

   The list in Section 3.1.1, "Period," entry 3, can be an exception: it's a sentence.

4. Use initial caps for the first word of each phrase, for example:

The prefetch operation consists of:
    Opcode recognition
    Instruction decode
    Address generation

If the words in the list are proper nouns, then, of course, use initial caps for each word in the phrase, for example:

The Prefetcher consists of three major units:
    Code Spotter
    Instruction Decoder
    Address Generator

5. Punctuate sentences in a list just as they would be punctuated in text.

6. Punctuate lists that form a sentence as you would a sentence in text, except use initial caps for each item in the list, for example:

    The Differential Position Signal:
        Keeps the heads on track,
        Detects cylinder crossing, and
        Monitors head speed across the cylinders.

7. When your introduction to the list mentions the number of items in the list and/or when the list contains four or more items, number each item in the list, for example:

    The tape contains the following four files:
        1. DOCUMENTATION
        2. SYSGEN
        3. DIAGNOSTICS
        4. REPORTS

8. When the list contains two or three levels, use numbers followed by numbers with single then double parens to differentiate the levels, for example:

The tape contains the following four files and their subfiles:

1. **DOCUMENTATION**
   1) *Installation Guide*
   2) *User's Guide*
2. **SYSGEN**
   1) **Parameter Table**
      (1) **Physical**
      (2) **Logical**
   2) **Initialization**
3. **DIAGNOSTICS**
4. **REPORTS**

9. If the list contains four levels, then begin the first level with a Roman numeral and repeat the above pattern one level down; that is:

   I. **DOCUMENTATION**
      1. *Installation Guide*
      2. *User's Guide*
  II. **SYSGEN**
      1. **Parameter Table**
         1) **Physical**
            (1) **Ports**
            (2) **Channel to channel adapters**
         2) **Logical**
            (1) **Real**
            (2) **Virtual**
      2. **Initialization**
         .
         .
         .

 III. **DIAGNOSTICS**
 IV. **REPORTS**

7
REFERENCES

Technical publications contain references to internal material and to external, supplementary, material.

Write and format all references consistently throughout the publication or complement of publications. Even if your readers are not conscious of patterns, the patterns help them to anticipate the relief of a figure or table that will follow or the direction to look for information necessary to continue a task.

All references should be double-checked before printing.

7.1 INTERNAL

Internal references pertain to figures, tables, sections, and appendixes in the same publication.

1. In general, use declarative sentences to refer to figures and tables, for example: **Figure 4-8 illustrates the process of chip design and fabrication**.

    NOTE: Place figures and tables as close as possible and always following their reference in text.

2. Use imperative sentences to refer to figures, tables, appendixes, or other sections that the reader must use in order to

accomplish a task in a procedure, for example: **See Appendix A for calculating Manhattan connections**.

3. When referring to other sections in the book, refer to the most helpful level:

    If the information appears only under the lowest level header, then refer to it precisely, for example: **Data filename conventions are discussed in Section 3.2.4.1**. If the header title is other than the same ("Data Filename Conventions" in the last example), then name the title also.

    If the information appears under two or more parallel level headers, then refer to the level header that contains them, for example: **"Filename conventions for each type of file are discussed in Section 3.2.4** (which is presumably titled "Filename Conventions" since its title is not referred to).

4. Avoid too many cross-references among sections, but if the material is too long to repeat, then tell the reader what to expect, for examples: **See codes that result from incorrect execution in Section 5.2, "Program Errors,"** or **After replacing the board, run the bring-up diagnostics described in Section 6**, or **The use of edit macros is discussed in Chapter 4**. (Some technical publications are divided into chapters rather than sections as first-level headers.)

    Do you see why the title of Section 5.2 in the foregoing examples is referred to? By the same rule, Section 6 must be titled "Bring-Up Diagnotics" and Chapter 4, "Edit Macros."

    NOTE: If you find you have to refer to information in other sections more than you'd like, then think about reorganizing the material.

5. If you don't have the capability for automatic numbering and renumbering, then refer to figures, tables, section levels, and appendixes by title rather than by number. Having to renumber references manually because of reorganization or revision is time consuming and fraught with chance for error.

6. Don't refer to page numbers. The references would have to be checked at the most hectic time of all.

7.2 EXTERNAL

External references point out other publications that are pre- or corequisite with your book or that contain supplemental information. External references must be listed in the preface and should be referred to where applicable in text.

7.2.1 In Bibliography

The bibliography of external references is placed in the preface under "REFERENCES" and almost always contains titles of other technical publications: publications in the complement, volumes in the publication, or publications about a product with which yours is compatible. Since the authors of technical publications are rarely cited, alphabetize the titles, for example:

> *Super-10 Installation Guide*, **1234565.**
> *Super-10 Principles of Operation*, **1234564.**

NOTE: Use part numbers but not revision levels.

Less frequently, you may need to list a book. In that case, use the standard bibliography form, for example:

> **Yarmish, Rina, and Joshua Yarmish.** *Assembly Language Fundamentals 360/370 OS/VS DOS/VS.* **Reading, Massachusetts: Addison-Wesley, 1979.**

1. Alphabetize the list of external references by title for publications and by first author's last name for books.
2. Don't number the list.
3. If you have the print capability, then set the titles in italics; if not, then use underlines.
4. Unless you really expect your reader to be familiar with or to read all of the referenced material, cite relevant chapters or sections, for example:

> *Super-10 Principles of Operation*, **1234564: Sections titled "Registers" and "Memories."**

7.2.2 In Text

In text, references to external publications are especially useful when you are describing a large system needing multivolume publications or when your company produces compatible machines and/or programs.

1. In general, use declarative sentences, for example: ***The CMOS Function Set Reference Manual* contains the list of parameters and their options.**
2. Use imperative sentences to refer to an external publication that the reader must use in order to accomplish a task in a procedure, for example: **If the green LED above the ON/ OFF button does not light, then see *The Micromecca X-23 Troubleshooting Guide*.**
3. If you have the print capability, set the publication titles in italics; if not, use underlines.
4. Don't refer to page numbers or headers of external publications. The reader can get the most accurate information from the referenced publication's table of contents.

8

APPENDIXES

Appendixes need not be included in all technical publications, but when they are, appendixes contain details that would break up the flow of information in a section, for example: long tables or series of computer instructions, error messages, and algorithms.

Appendix format follows the rules according to content; for example, if you are presenting error messages, then use error message format. However, all appendixes are subject to the following considerations:

1. Appendixes follow the last section of the publication, beginning on right-hand pages.
2. Place multiple appendixes in the same order in which they are first referred to in text.
3. Identify each appendix by a capital letter and its title, for example:
   APPENDIX A—PRECOMPILER ERROR MESSAGES
   The running foot should carry the same information.

9

GLOSSARY

Every technical publication needs a glossary. Technical terms vary from discipline to discipline and from company to company. Each publication contains terms specific to the product. Though a term may be defined when it is first used in text, readers don't necessarily read the publication in sequence, nor are they pleased about having to search backward for the first mention of an acronym.

Short glossaries (a dozen or fewer terms) may be included in the introduction. A longer glossary follows the appendixes and begins on a right-hand page.

Glossaries are due the following considerations:

1. Include all acronyms used in the publication. Define each by its spelled out form.
2. Include all proper nouns used in the publication. Define each in terms of physical and/or functional characteristics.
3. Include all terms that you define or explain in the publication. These terms vary greatly depending on the purpose and audience. You certainly wouldn't want to gloss *read* in a programmer's guide, but might well do so in a guide meant for new data processing operators.

4. Use only the most common acronyms in the definition of a term, for instance: *CPU* and *I/O*.

5. If you refer to other glossary entries, then include the reason for their relevance; for example after the definition of *call*: **See *return* for subroutine return options.**

6. Capitalize only those entry names that are capitalized in text.

7. Present all entry definitions in parallel grammatical structure. Use a phrase beginning with *A*, *An*, *The*, or a gerund to begin the definition of a singular noun; use a verbal form (the same one) to begin the definition of a verb.

   You may follow a definition by further explanation in sentences. For example:

   > *Datadump.* **The utility that prints unformatted data to the printer. Datadumps are used for debugging internal database record labels.**

8. Alphabetize all entries by the dictionary method rather than the telephone book method, that is, by letter rather than by word. For example:

   > *Data Bus Interface.* **The functional unit that controls other units' access to the Data Bus.**
   > *Data Code Table.* **The system table that controls user access to data. Only users with a code of 4 or above can access the table.**
   > *DBI.* **Data Bus Interface**
   > *DCI.* **Data Control Table**

9. Identify the glossary by the title "GLOSSARY." The running foot should carry the same information.

10

INDEX

An index is an alphabetic list of significant subjects and the pages on which they are discussed in the publication. An index follows the glossary and begins on a right-hand page. Indexes are due the following considerations:

1. Include two or three ways for your reader to access a topic.
   1) Software book indexes should provide both feature-oriented entries (the commands) and task-oriented entries (the utilities), for example:

      PRINT Command, 21
      Printing
      Program listings, 21
      Text files, 22−23

   2) Hardware book indexes should provide both assembly-oriented entries (the parts) and task-oriented entries (the procedures), for example:

      Blower Assembly, 160
      .
      .
      .
      Replacing the blower, 164

2. Include all acronyms. If the term was used enough to require an acronym, then it is important enough to index. List only the page on which the acronym was first used.

3. Limit an entry to three levels, beginning with the most general topic as the first-level entry, for example:

database
  interface with system, 72
  records, 43–47
    record fields, 48–63
    structure of, 2–5

NOTE: Unlike sections and subsections, there need not be a minimum of two subentries to a first- or second-level entry.

4. Alphabetize parallel level entries by the dictionary method rather than the telephone book method, that is, by letter rather than by word, for example:

record format, 45
recording, 21
records, 43

5. Identify the index by the title "INDEX." The running feet should carry the same information.

Because of the user-unfriendliness of the automated indexing systems that I'm aware of and the tremendous time and chance for error involved in manual creation of indexes, many publications do not have helpful indexes—or any indexes at all.

If you do have the personnel—usually a work study or junior employee—to create an index manual, then use index cards. After skimming through the publication once, the index cards should contain all the planned first-level entries. When you've been through the publication the number of times you have time for, the index cards should contain all the entries and their page numbers. Once the index cards are complete, shuffle them alphabetically and transfer the information to the index.

Of course you can't make the final pass at the index until the book is in production—another reason that indexes are uncommon.

If you do not have the mechanical aids or personnel to include indexes in your books, then be sure that the table of contents lists titles through the third level.

Did you notice that the information in the last sentence is discussed both under "Table of Contents" in Chapter 1 and here?

PART II: LOGICAL COMPONENTS

Technical writing, like all technical disciplines, is built on basic logical structures. Just as it is assumed that an engineer understands material properties, that a systems programmer understands console modes, and that a field service rep understands the diagnostic functions, a competent hightech writer needs to understand rhetorical techniques in order to achieve the most elegant and proper solutions.

Some writers assume that the audience does not recognize inappropriate rhetorical conventions, that users are unaware of incorrect punctuation or illogical connections; however, some readers of technical publications are also avid readers in other fields. The other readers, probably the majority, may never recognize a grammatic slip, but they shouldn't be expected to understand prose that is more complicated than the electronics being explained.

Good technical writing is like a good system architecture. Implementations are transparent to the user; that is, the writing simplifies complex subjects.

A good technical writer is not only a good writer, a writer who is concerned with grammar, who understands how the structures of written English relate, where to place adverbs, when to choose *continual* over *continuous*; but a good technical writer is also a writer who uses the specialized diction, definition, and descriptive techniques particular to technical rhetoric.

In addition, technical publications are often translated into some languages deceptively like and others obviously unlike English. Not many translators are truly bilingual, so the technical writer can help by using clear, consistent logical structures to show cause, coordination, differentiation, sequence. It'll help the English speakers too.

Three comma rules are exemplified in the next-to-last sentence in the above paragraph. Did you notice that there is no comma between "consistent" and "logical"? If you are unsure of the reason, then see comma rule 6 in Section 3.1.2. If you are unsure of which rules were used, then see all the comma rules in Section 3.1.2.

This part of the book is the most complex because, as with all metalogical systems, written English must be discussed in terms of itself.

11
GRAMMAR

A written language is a set of alphabetic symbols and a set of rules about ordering those symbols to make statements. When a computer is presented with a statement, it performs one or a sequence of its functions. Natural language, for example, English, contains much larger sets of symbols and ordering rules than any computer language.

Grammar is the systematic way in which the set of ordering rules manipulates symbols to form structural patterns that convey meaning, the way in which parts of speech form a coherent system of communication.

Grammar rules come from grammarians, who may be dubbed "Custodians of the Language" on one side or "Advocates of Usage" on the other. The Custodians say that the old rules should determine how people speak. The Advocates say that how people speak now should determine the rules. Since this book is about written language, and of a type certainly requiring precision, I tend toward being a Custodian; however, in some instances, Section 12.4, "Gender," for example, I'd be termed an Advocate, so much so that I will accept disagreement between two pronouns (as in "Everyone should first define their program interfaces)" rather than accepting a masculine or feminine pronoun, even though the publisher disagrees.

Grammarians will always disagree, so learn enough about grammar to make your own decisions because understanding grammar, its usage and structure, is basic to technical writing.

If you are unsure of the explanations in this chapter, or if you are interested in pursuing the fascinating nuances of grammar beyond this scratch on its surface, then one or more of the entries in the bibliography will be a good place to start.

11.1   TERMS

Like any discipline, grammar needs a specific body of terminology to describe its structures as precisely as possible. The following alphabetic list of grammar terms is not exhaustive but meant to be a reminder. Unless you have recently taken a grammar course (or taught one), skim the terms before going on and refer to them as needed as you read this book. Remember, **bold** indicates good examples; [ ] indicates bad.

ADJECTIVE - *Describes or limits a noun or noun phrase*

For examples: In **the long, complicated instructions** and **The instructions are long and complicated**, "long" and "complicated" are descriptive. In **the first five instructions**, "five" is limiting.

What is "first"? If you are unsure, then the next entry in this list will help.

ADVERB - *Describes a verb, adjective, or another adverb*

An adverb tells where, when, how, or to what extent. For examples: In **Move the bolt forward**, "forward" tells where; in **Next, move the bolt**, "next" tells when; in **Move the bolt cautiously**, "cautiously" tells how; in **Move the bolt slightly**, "slightly" tells to what extent.

In the above examples, the adverbs modify verbs, but they may also modify adjectives or other adverbs. For example, in **Move the very heavy casing very carefully**, the adverb "very" modifies the adjective "heavy" and the adverb "carefully."

Many adverbs are formed from an adjective suffixed by *-ly*, for example: *considerably, significantly, slowly, quickly*.

ANTECEDENT - *A word to which a pronoun refers, generally the subject, sometimes the direct object, of the sentence*

If a pronoun matches both the subject and object, then it defaults to the subject. For example, in **Place the cursor in the center of the circle, and move it to the upper left on the screen**, "cursor" (not "circle") is the grammatic antecedent of "it."

APPOSITIVE - *A noun or noun phrase that immediately follows and restates the preceding noun or noun phrase*

For examples, in **Run Early Placement and Final Placement, the first two programs, as one job**, "the first two programs," is in apposition to "Early Placement and Final Placement"; in **Two components, the cache and the I/O structure, determine the price/performance balance characteristics**, "the cache and the I/O structure" is in apposition to "Two components."

ARTICLE - *A subclass of adjectives consisting of three words:* a, an, *and* the

A and *an* are the indefinite articles, which denote unspecified items, for example: **A program runs faster in batch mode than in interactive mode**, meaning any program.

The choice between *a* and *an* depends on the spoken word. Use *a* before words and acronyms beginning with a consonant sound and *an* before words beginning with a vowel sound, for examples: **a half hour, a CPU, an hour, an I/O device**.

The is the definite article, which denotes an already specified item, for example: **The program takes 20 seconds of CPU time**, meaning a specific program. *The* may also be used to introduce a topic, for example: **The metalization process requires five major steps**.

CLAUSE - *A group of words containing a subject and verb*

An *independent clause* can stand alone as a sentence; a *dependent clause* (also called *subordinate clause*) must be connected to an independent clause. For example, in **Pull the power switch to OFF before you open the cabinet**, "pull the power switch to OFF" is an independent clause; the rest of the sentence is a dependent clause, which cannot stand alone.

CONJUNCTION - *A word that joins two or more sentence elements*

Conjunctions are adverbial, coordinating, correlative, or subordinating.

ADVERBIAL CONJUNCTIONS begin and/or join independent clauses and include: *also, however, instead, moreover, nevertheless, otherwise, still, then, therefore,* and *thus*; also, they may be called *conjunctive adverbs*. For instance, "also" is an adverbial conjunction in the previous sentence.

NOTE: Adverbial conjunctions that connect independent clauses are preceded by a semicolon and followed by a comma.

COORDINATING CONJUNCTIONS join two or more words, phrases, or clauses that are logically and grammatically parallel. There are only seven coordinating conjunctions: *and, or, for, nor, but, yet,* and *so*. For instance "and" is the coordinating conjunction in the previous two sentences, joining parallel adverbs ("logically," and "grammatically") in the first sentence and joining items in a series (the coordinating conjunctions) in the second; "or" is the coordinating conjunction in the first sentence, joining two adjectives ("two" and "more") and joining items in a series ("words," "phrases," and "clauses").

CORRELATIVE CONJUNCTIONS are used in pairs to join two or more words, phrases, or clauses that are logically and grammatically parallel. There are six correlative conjunctions: *as . . . as; both . . . and; either . . . or; neither . . . nor; not only . . . but also; whether . . . or.* You must use not only the first but also the second member in each pair. For instance, "not only . . . but also" is the correlative conjunction in the previous sentence.

SUBORDINATING CONJUNCTIONS join dependent clauses to independent clauses; in fact the subordinating conjunctions make the clauses dependent. The following are common subordinating conjunctions: *although, after, as, because, before, if, unless, until, when*. When you use subordinating conjunctions, be sure to punctuate the sentence correctly. For instance, "when" is the subordinating conjunction in the previous sentence. If you are unsure of the punctuation of an introductory dependent clause, then see comma rule 3 in Section 3.1.2.

EXPLETIVE - *A word that fills the position of the grammatical subject of a clause*

The two expletives are *it* and *there*. For examples, in **It is necessary to run the diagnostics**, "It" is the expletive; in **There are six files on tape**, "There" is the expletive. Expletives also include exclamations and obscenities—but not in tech writing.

English has the interesting property of requiring *it* in constructions describing weather conditions, for instance: It's raining.

GERUND - *A verbal that ends in* -ing *and is used as a noun*

For example, in **Programming is easier on the Super-10**, "Programming" is the gerund. Like other nouns, gerunds may head phrases, for example: **Programming on the Super-10 is easy**.

NOTE: Don't confuse a gerund with a present participle modifier or the present progressive tense.

INFINITIVE - *A verbal that is preceded by* to *and is used as a noun*

For example, in **The Super-10 is easy to program**, "to program" is the infinitive. Like other nouns, infinitives may head phrases, for example: **To program the Super-10, use Super-basic**.

MODIFIER - *Words, phrases, and clauses that expand or limit the meaning of a subject, verb, or direct object*

For example, in **The Super-speed option increases program throughput dramatically by allowing asynchronous processing and channel control**, "option" is the subject, "increases" is the verb, "throughput" is the direct object, and the rest of the sentence is modification.

Modification, including adjectives, adverbs, appositives, relative clauses, and verbal and prepositional phrases, is discussed in Section 11.3.1.2.

NOUN - *The name of classes, people, places, things, and abstractions*

Nouns are common or proper, count, mass, or collective.

COMMON NOUNS name a general class, person, place, thing, or abstraction, for examples: **computers**, **systems analyst**, **electronics company**, **chips**, **perseverance**.

PROPER NOUNS title a specific class, person, place, thing, or abstraction and are capitalized, for examples: **Micromecca Systems**, **Dr. Gene Amdahl**, **Chairman of the Board**, **Micromecca Corporation**, **Super-10 Chip**.

Can you think of a case of an abstract noun that should be capitalized in technical writing? I can't.

In technical writing, proper nouns are most often used for the name and parts of the product. Capitalize job titles when they are tied to the name of a particular person, but consider them common nouns when they name a job title alone, for examples:

> **Dr. Gene Amdahl, Chairman of the Board of Trilogy Limited, also started up Amdahl Corporation.**
> **The systems analyst who plans new systems must be aware of the hardware, software, and people involved.**

COUNT NOUNS identify things that can be separated into countable units, for examples: **instructions**, **pin locations**, **bolts**, or **ideas**.

MASS NOUNS identify things that compose an uncountable

amount that can be measured, but not separated, for examples: **electricity** and **silicon**.

COLLECTIVES, a kind of count/mass noun that sometimes causes subject-verb puzzlement, are discussed in Section 11.2.1.2.

After reading "Collectives" and rule 4 in Section 11.3.1.2, "Relative Clauses," come back to the last sentence above and decide whether you would change "causes" (the verb of "noun") to plural. Consider whether "collectives" and "noun" qualify as collectives.

NUMBER - *A grammatical property of nouns, pronouns, and verbs indicating whether they are singular or plural*

Most nouns form the plural by adding *s*, for example: **computer** → **computers**. Nouns that end in *s* or *x* generally add *es*, for example: **radix** → **radixes**.

For irregular plurals (generally latin forms), you may want to consult a dictionary, but once you've chosen, be consistent: Always use the form you choose, for example: *zeros* over *zeroes*. If you chose *appendixes* over *appendices*, choose *indexes* over *indices*.

All personal pronouns except *you* change form in the plural:

I → *we*
he, she, it → *they*
me → *us*
him, her, it → *them*

Most verbs express the third person singular by *s*. For example, compare **The initialization program runs first** with **The programs run in sequence**.

OBJECT - *A noun or pronoun that takes the action of a verb or ends a prepositional phrase*

For examples: in **Remove the rotor cap** and **Remove the cap from the rotor**, "cap" is the object of the verb "remove."

However, "rotor" is an adjective in the first example and the object of the preposition "from" in the second.

PARTICIPLE - *A verbal that is used as an adjective*

Like all adjectives, participles describe nouns. A present participle ends in *-ing*, for example: **Programming languages are proliferating**.

NOTE: Don't confuse a present participle with a gerund or present progressive tense.

Most past participles end in *-ed*, but may end in *-d*, *-t*, *-n*, *-en*, or be formed by a vowel change in the verb, for examples: ended, held, kept, known, fallen.

Did you happen to notice that the examples followed the exact sequence as the past participle endings ("ended" for "-ed," "held" for "-d," "kept" for "-t," and so forth)?

PARTS OF SPEECH - *The classification of words according to their function in a sentence*

If a word names something in a sentence, then the word is a noun in that sentence; if a word shows the action or the state of being of the subject of a sentence, then it is a verb in that sentence; if a word describes another word in a sentence, then it is an adjective or adverb in that sentence; if a word shows relationships among elements in a sentence, then it is a conjunction or a preposition in that sentence.

For example, in **Program the computer**, "Program" is the verb, and "computer" is a noun; in **Computer programs are proliferating**, "programs" is a noun, "computer" is an adjective, and "are proliferating" is the verb (present progressive).

PASSIVE - *See* VOICE

PERSON - *The grammatical point of view: first, second, or third*

First person—the speaker or writer (*I* or *me*, and *we* or *us*);

Second person—the hearer or reader *(you)*

Third person—the person or thing written about (*it, they, them*, and all nouns)

First person is rare in technical writing. Second person is generally imperative without the word *you* and describes steps to be performed; third person generally describes the systems.

In case you've noticed the lack of *he* and *she*, its because they're unnecessary in technical writing. Section 12.4, "Gender," talks about alternatives.

PHRASE - *A group of words that has no subject, no verb, or neither*

Phrases are gerundial, infinitive, participial, prepositional, noun, or verb.

GERUNDIAL PHRASES contain a gerund and function as nouns; for example, in **Repairing the machine consists of merely removing one card and replacing it with another**, "Repairing the machine," "merely removing one card," and "replacing it" are gerundial phrases.

INFINITIVE PHRASES contain an infinitive and function as nouns; for example in **It is unnecessary to diagnose the problem beyond board level**, "to diagnose the problem" is an infinitive phrase.

PARTICIPIAL PHRASES contain a participle and function as adjectives; for example, in **Well known in the industry, the D-algorithm is often used**, "Well known" is the participial phrase.

PREPOSITIONAL PHRASES begin with a preposition, end with a noun or pronoun, and function as adjectives or adverbs; for example, in **The drives are located on the bottom of the card cage**, "on the bottom" and "of the card cage" are prepositional phrases.

NOUN PHRASES or VERB PHRASES contain a noun or verb and

its modifiers; for example, in **The functional simulator quickly performs logic calculations**, "The functional simulator" is a noun phrase (subject); "quickly performs" is a verb phrase; and "logic calculations" is a noun phrase (direct object).

PREPOSITION - *A word that shows relationship between two elements in a sentence and begins a phrase that ends in a noun or pronoun*

The following words often function as prepositions: *about, above, against, among, at, before, behind, between, by, for, from, in, of, on, onto, to, toward, under, with, without.*

However, many words that function as prepositions also function as other parts of speech. For examples: in **Remove the microprocessor chip on the Channel Card**, the "on" functions as a preposition; in **Turn on the switch**, the "on" functions as the second part of a two-part verb (even if the sentence is written **Turn the switch on**); in **Turn the switch to ON position**, "ON" functions as an adjective; in **Turn the switch to ON**, "ON" functions as a noun, the object of the preposition "to."

PROGRESSIVE - *An* -ing *form of the verb preceded by* to be *in the appropriate tense*

A progressive verb can happen in any tense. For example:

present	The system *is running*.
present perfect	It *has been running* for three months straight.
past	It *was running* payroll yesterday.
past perfect	It *had been running* routing programs before that.
future	It *will be running* routing programs again tomorrow.
future perfect	In one more month it *will have been running* longer than any system in the shop.

PRONOUN - *A word used to substitute for a noun*

A pronoun must match its antecedent. For example, in the previous sentence "its" is the pronoun, and "pronoun" is its antecedent.

Pronouns are demonstrative, indefinite, personal, or relative.

DEMONSTRATIVE PRONOUNS—*This*, *that*, *these*, and *those*—are correct in technical prose only before a noun.

INDEFINITE PRONOUNS specify a class. Some common indefinites are: *all, any, both, each, either, few, many, most, neither, none, several, some*. For example, in **The programs are in a jobnet so that all begin automatically**, "all" is the indefinite pronoun referring to "programs."

PERSONAL PRONOUNS are limited almost exclusively to third person in technical writing, and should be confined to *it, its, they, them*, and *their*; the section on gender, Section 12.4, provides the rationale. You might state second person (*you, your*) occasionally.

RELATIVE PRONOUNS—*that, what, which, who, whom*, and *whose*—begin a relative clause, which is a type of dependent clause. In the previous sentence, "which is a type of dependent clause" is an example of a relative clause. Relative clauses are discussed in Section 11.3.1.2, "Modification."

SENTENCE - *At least one independent clause that begins with a capital letter and ends with a period*

Sentences are declarative, imperative, interrogative, subjunctive, and exclamative in type.

DECLARATIVE SENTENCES (also called indicative) define and describe, for example: **The system runs diagnostics first**.

IMPERATIVE SENTENCES direct, for example: **Run system diagnostics first**.

INTERROGATIVE SENTENCES question, for example: **Does the machine run diagnostics first?**

SUBJUNCTIVE SENTENCES express matters not realized, for example: **If the machine were to run diagnostics first, then most bring-up problems could be avoided.**

NOTE: The absence of the *s* for third person verbs and the use of *be* and *were* express subjunctive in English.

EXCLAMATIVE SENTENCES, which end in exclamation points, are best avoided in technical writing—even in marketing publications.

Sentence structure—*simple*, *compound*, *complex*, and *compound-complex*—is discussed in Section 11.3.1, "Sentences."

SUBJECT - *A word or group of words that performs or receives the action of the verb or is described by the verb*

For examples, in **The CPU performs at 5 MIPs**, "CPU" is the subject, and in **The CPU is a 5-MIP machine**, "CPU" is the subject, but in **Five MIPS can be achieved by the CPU**, "MIPS" is the subject.

What is "CPU" in the third example? Does *object of a preposition* or *passive voice agent* explain its function? If you are unsure, then look under "VOICE."

TENSE - *Grammatic time markers on verbs*

Following are the three basic tenses and their perfect forms:

Present: The control program *begins* with the initialization sequence.

Present perfect: The sequence *has begun*.

Past: It *began*.

Past perfect: It *had begun* without operator intervention.

Future: It *will end* when all of the checking routines produce a positive condition code.

Future perfect: It *will have run* 843 independent routines at the end of 30 seconds.

VERB - *A word or group of words that name an action or a state of existence*

Verbs are transitive, intransitive, or linking.

TRANSITIVE VERBS require a direct object. For example, in **They installed the system in four hours**, "installed" is the transitive verb, and "system" is the direct object. ("They" is the subject, and the rest of the sentence is modification.)

INTRANSITIVE VERBS don't take an object. For example, in **The system ran for 8573 hours without a failure**, "ran" is the intransitive verb, and "system" is the subject. (The rest of the sentence is modification.)

NOTE: *Install* is always transitive (whether active or passive voice), but *run* can be transitive as well as intransitive, for example in **The machine runs the UNIX Operating System**, where "System" is the direct object.

LINKING VERBS are generally some form of *to be*. For example, in **The system is up**, "is" links the subject, "system," to its adjective, "up." *Appear*, *seem*, and *become* are also linking verbs.

If you are interested in verbs, linking verbs especially, then there are tons of papers and books that you can get to expand the topic into inchoates, statives, complementaries, and many other beguiling classifications beyond this mere mention.

VERBAL - *A derivation of a verb used as a noun, adjective, or adverb; also called nonfinite verbs*

There are only three verbals: gerund, infinitive, and participle. They are discussed individually under their own titles, collectively under "PHRASE," and in Section 11.3.1.2, "Modification."

VOICE - *The two forms of a clause that indicate whether the grammatical subject either performs the action (active voice) or is acted upon (passive voice)*

For examples: In **The programs in the jobnet run simulation routines**, "programs" is the agent, performing the action of the verb "run," so the sentence is active. In **The programs in the jobnet are run by batch processes**, "programs" is the patient, receiving the action of the verb "are run," so the sentence is passive.

Two *passive markers* signal passive voice:

1. The passive sentence contains a form of *to be* plus a past participle; for example, in **The operand is stored in the Operand2 Register**, "is stored" is the passive verb form.

NOTE: If a sentence is both passive and progressive, then two forms of *to be* are needed; for example, in **While the instruction completes, the next instruction's pre-fetched operand is being stored temporarily in the operand stack**, "is" is passive and "being" is progressive.

2. The passive sentence often contains a phrase beginning with *by* and ending with the *agent* of the action, for example: **The machine is booted up by console software**. However, the agent of the action need not be stated in order for a sentence to be passive; for example, in **The machine is booted up after each scheduled or unscheduled maintenance**, "machine" is the *patient* (object) of the action, and the *agent* is not stated.

11.2   RULES

Grammar usage rules come in two forms: prescriptive and proscriptive. Prescriptive rules tell you what to do; proscriptive, what not to do. This section contains both kinds of rules, but deals only with common grammar confusions—subject-verb agreement, pronoun reference, and verb forms—there are lots more rules.

11.2.1   Subject-Verb Agreement

A verb must agree with its subject.

Don't let intervening clauses or phrases obstruct your view of the subject with which the verb must agree. For example, in **The set of applications that run circuit design, placement, and routing routines is based on university research programs**, "set" is the subject of the sentence and of the verb "is based on" ("run" is the verb of the relative clause).

11.2.1.1 *Amounts*

Amounts include terms that define sums, rates, and quantities. Amounts generally take a singular verb, for examples: **No more than 20 hours of maintenance is required** and **180 ps is the switching rate**.

11.2.1.2 *Collectives*

A collective is a noun that groups discrete units. Its verb may be singular or plural dependent on whether you consider it a mass or count noun in the context of the sentence.

If you want to stress unity, then use a singular verb form, for example: **The group of utility programs provides file maintenance.**

If you want to stress individuality, then use a plural verb form, for example: **The group of utility programs provide file opening, closing, and deleting functions.**

What is the subject of both example sentences? If you said "group," then you're right. What's "programs"?

"Data" is another example of a collective. If you want to stress unity (which is generally the case) then it takes a singular verb form, for example: **The data is stored in Register11 and Register12**. If you want to stress individuality, then it takes a plural verb form, for example: **The data are transmitted serially**. If you switch from a singular to plural verb, have a good reason. (Some academic writing still uses *data* and *datum* to distinguish plural from singular.)

11.2.1.3 Compound Subjects

Compound subjects are linked by the coordinating conjunctions *and* and *or* and by correlative conjunctions.

A subject containing *and* takes a plural verb, for example: **The power indicators and the power have failed**.

A subject containing *or* takes either a singular or plural verb, depending on the second item in the pair. For example, compare **The power indicators or the power has failed** with **The power or the power indicators have failed**.

A subject containing *as . . . as*, and *both . . . and* and *not only . . . but also* takes a plural verb, for example: **Not only the motor generator but also the power supply have backups in case of failure**.

A subject containing *either . . . or*, *neither . . . nor*, or *whether . . . or* takes a singular or plural verb, dependent on the second item in the pair. For example, compare **Either the connectors or the motor generator needs replacement** with **Either the motor generator or the connectors need replacement**.

11.2.1.4 Indefinites

When a subject is an indefinite pronoun, the verb may be singular or plural depending on the pronoun's antecedent. *Both*, *few*, *many*, *most*, and *several* take plural verbs, for example: **The programs are in a jobnet, so several automatically run in sequence**.

Any, *another*, *each*, *either*, *neither*, and *none* generally take singular verbs, for example: **The programs are in a jobnet, so none requires operator intervention**.

All and *some* depend on whether the antecedent is a mass or count noun, for examples: **Some (oil) is leaking and All (connections) are soldered**.

If an indefinite subject is followed by a prepositional phrase, then subject-verb agreement is clearer, for examples: **Some of the power units require replacement** and **All of the coolant requires draining**.

11.2.1.5 *Subjunctive*

In a subjunctive sentence, *were* agrees with all subjects, both singular and plural, for example: **If an artificial intelligence machine were to be built, then theories could be tested.**

11.2.2 Pronoun Reference

Pronoun reference is especially important in technical writing because a pronoun carries no information except by reference to its antecedent.

When there is doubt about the antecedent, a personal pronoun grammatically defaults to the subject of a clause, but often the pronoun seems to refer to the last-mentioned noun. For example, in [The input consists of the vector chain; it includes logic and arithmetic checks], the "it" grammatically refers to "input"; however, it might be read to refer to "chain." If there is any doubt, then repeat the antecedent.

Use the demonstrative pronouns *this* and *that* only with a noun following. For example, in [Blocked data sets minimize disk space and reduce resource use; this will improve a program's performance], what does the "this" refer to? Repeating the antecedent ("blocking data sets") is one way to solve the problem. You might also recast the sentence by putting it into imperative: **To improve program performance, block data sets to minimize disk space and resource use**.

Former and *latter* often serve as pronouns because they refer to previous information; use them sparingly and only to refer to two items.

11.2.3 Verb Forms

The following verb forms seem to confuse some technical writers and may confuse many readers.

11.2.3.1 Comprise

Because many dictionaries define the verb *comprise* as "to constitute parts of a whole" and as its opposite, "to contain a whole of parts," either pick one definition and use *comprise* consistently, or, better yet, don't use *comprise* at all.

11.2.3.2 Due To

If you use *due to* mean "because of," then use it only after a form of the verb *to be*. For example: **The program failure was due to a system bug** is correct; [The program failed due to an operator error] is incorrect. Replacing "due to" with "because of" is even better: **The program failed because of a system bug**.

11.2.3.3 Got/Gotten

Both *got* and *gotten* are proper past participles of the verb *to get*, but *gotten* connotes a progressive movement, and *got* connotes a finished action, for examples: **Operating systems have gotten more specialized** and **Once programs have got their data files, processing can begin**.

NOTE: *Has got* is not a synonym for *must* or *has*. For example: [The program has got a bug] should be written without the "got"; and [you have got to specify all device parameters] should be written "you must . . . ," or better yet, **Specify all device parameters**.

11.2.3.4 Lie/Lay

The difference between *lie* and *lay* is that *lie* is intransitive and *lay* is transitive.

Lie does not take an object, for example: **If the floppy lies in the sun, then the data will be destroyed**.

The third person singular and plural forms of lie are:

present:	lies, lie
present progressive:	is lying, are lying
present perfect:	has lain, have lain
present perfect progressive:	has been lying, have been lying
past:	lay
past progressive:	was lying; were laying
past perfect:	had lain
past perfect progressive:	had been lying
future:	will lie
future progressive:	will be lying
future perfect:	will have lain
future perfect progressive:	will have been lying

Lay takes an object; for example, in **The robot lays a full cradle of wafers into the photolithography tray**, "cradle" is the object.

Technical writing most often uses the second person present form (imperative) of *lay*, for example: **Lay the cable between the plug and the power unit**.

The second person forms of lay are:

present:	lay (the object)
present progressive:	are laying (the object)
present perfect:	have laid (the object)
present perfect progressive:	have been laying (the object)
past:	laid (the object)
past progressive:	were laying (the object)
past perfect:	had laid (the object)
past perfect progressive:	had been laying (the object)
future:	will lay (the object)
future progressive:	will be laying (the object)
future perfect:	will have laid (the object)
future perfect progressive:	will have been laying (the object)

11.2.3.5 Sit/Set

The difference between *sit* and *set* is that *sit* is intransitive and *set* is transitive.

Sit does not take an object, for example: **The terminal sits on a small desk**.

The past and past perfect tenses of *sit* are the same: sat.

Would using "is" instead of "are" in the previous sentence be correct? Try Section 11.2.1.2, "Collectives"; could you say that "tense" is a collective? While we're at it, what are "past" and "past perfect" in that sentence? (Try adjectives.) What is "past and past perfect tenses"? (Try noun phrase.)

Set takes an object. For example, in **The switches must be set to OFF**, "switches" is the object (the patient in a passive voice sentence).

Technical writing most often uses the second person (imperative) form of *set*, for example: **Set the registers to zero**.

The past tense and present perfect tense of *set* are also *set*.

****"Are" or "is"? What did you decide? See the second paragraph under Section 11.2.1.3, "Compound Subjects." How about if you put an *s* at the end of "tense"?***

11.3   STRUCTURES

The grammatical structures discussed in this section include sentences, paragraphs, transitions, and parallelism; parallelism is the overall technical writing structure.

11.3.1   Sentences

The sentence types—declarative, imperative, interrogative, subjunctive, and exclamatory ("SENTENCE" in Section 11.1 "Terms")—can be structured in four ways: *simple*, *compound*, *complex*, and *compound-complex*.

SIMPLE - *One independent clause*

Examples:

Computers run programs.
Programs run.
Press the ON switch.
The ATTENTION switch, a momentary-on switch, initiates the rezeroing operation by moving the heads to cylinder 0, resetting the address register, and sending the Ready signal to the controller.

COMPOUND - *Two or more independent clauses joined by a coordinating conjunction or a semicolon*

Examples:

The keyed-in data disappears from the screen, and a prompt message appears.
First replace the blower power plug; then check air pressure.

COMPLEX - *One independent clause and one or more dependent clauses in any order*

Example:

When the ROM has been checked and verified, the program returns to the initiation routine, which then calls for a RAM check.

COMPOUND COMPLEX - *Two or more independent clauses and one or more dependent clauses in any order*

Example:

Remove AC power cables and allow at least 15 seconds for the capacitors to drain before performing maintenance on the AC power distribution or filter-rectifier assemblies.

Try taking the above example sentences apart: decompose them into subject, verb, clauses, phrases. Are the clauses active or passive voice? Are the sentences declarative or imperative? If you have any doubts as to which structure is performing which function, then books listed in the first two sections of the bibliography will help.

In addition to the five types and the four structures, sentences come in two stylistic structures: *loose* and *periodic*.

A loose sentence begins with an independent clause followed by dependent clauses, modification, or both, for example: **The microprogram verifies the data registers by sending out first all 0s and then all 1s to two different registers, which are read back and compared each time**.

A periodic sentence is just the opposite: A periodic sentence ends with an independent clause preceded by dependent clauses or modifiers or both, for example: **When you have completed the data files, constructed the simulator, run the simulator against the data, and received the reports, look at the reports for any errors.**

11.3.1.1   Coordination and Subordination

Coordination and subordination are formal grammatical matters: Coordinating conjunctions join independent clauses; subordinating conjunctions begin dependent clauses.

Choose the correct conjunction to make the logic of the sentence clear. For example, *and* is never a logical substitute for *because*: [The power failed and the system shut down] is not as clearly causative as **Because the power failed, the system shut down**. To show coordinating rather than subordinating cause, use the coordinating conjunctions *so* or *for*, for examples: **The power failed, so the system shut down** and **The system shut down for the power failed** (a little archaic).

A compound sentence should be balanced by clauses that are of similar length and construction, and both clauses should show ideas of similar content and importance. The first sentence in this paragraph is an example. For another example: **Run the signal cable up the bracket side of the frame, but do not connect the cable yet**.

****"Yet" is not a coordinating conjunction in the example immediately above, so what is it? Try "ADVERBS" in Section 11.1, "Terms."***

Grammarians have traditionally said the important idea is supposed to be placed in the independent clause of a complex sentence.

When you make any kind of logical connection by means of a complex sentence, be sure you subordinate the correct clause—correct according to what you mean to say. For example, compare **Although the program uses less CPU time, the elapsed time is longer** with **The program uses less CPU time although the elapsed time is longer**. The first example stresses that the elapsed time is longer, and the second example stresses that the CPU time is shorter.

However, the idea in a dependent clause is equally important in the cause-and-effect statements so common to technical writing. If you are making a conditional cause and effect statement, then cast the sentence in an *If . . . , then . . .* structure. The previous sentence is an example. For another example: **If the VAT bit is on, then virtual address translation is in effect**.

11.3.1.2 Modification

Adjectives, adverbs, appositives, relative clauses, and prepositional and participial phrases—all the parts of the sentence except the subject, verb, and object of the clauses—are modifiers.

Some grammarians view dependent clauses as modification also, but because two of the four sentence structures require dependent clauses and because complex sentences form the largest part of technical text, only dependent relative clauses are included here; dependent clauses are discussed in Section 11.3.1, "Sentences."

Following are examples of modification in a sentence where "cable" is the subject and "runs" is the verb.

> **The** (article) **200-signal** (adjective) **cable, a flat coax** (appositive), **runs directly** (adverb) **to the controller** (preposi-

tional phrase), **which can be placed no more than ten feet away** (relative clause).

Modifiers—adjectives, adverbs, appositives, relative clauses, prepositional phrases, and participial phrases—should be placed as close to the word they modify as is possible. For examples: **Inputs are changed at each clock cycle independent of circuit behavior specified by STEP commands** says that STEP commands specify circuit behavior. **Inputs specified by STEP commands are changed at each clock cycle independent of circuit behavior** says that STEP commands specify input. Either could be a true statement, but only one is true of the operation being described.

1. *Adjectives.*   Technical writing abounds with adjectives because of the need for precise modification.

Nouns are often used as adjectives, but too many nouns strung together causes ambiguity, making the reader strain to figure out what is modifying what. For example, in [Latch input pattern value changes occur at each clock cycle,] the noun that is the subject ("changes") is preceded by four—count 'em, four—other nouns. To complicate matters more, "changes" can function as a plural noun or as a singular verb.

Better alternatives to stacking adjectives exist, for examples, recasting the sentence by adding a few prepositions: **Values at latch inputs change patterns at each clock cycle** and by adding hyphens: **Latch input-pattern values change at each clock cycle**.

You can usually move stacked modifiers to prepositional phrases following the noun; for example, "top air filter guard" could become "top guard for the air filter," but [guard for the top of the air filter], which contains too many prepositional phrases, is just as bad as too many stacked nouns.

Often inserting a hyphen will help resolve ambiguities in modification; for example, an "electric-panel opener" is a tool that opens panels that work electrically, and an "electric panel-opener" is a tool that works electrically to open panels.

2. *Adverbs.*   An adverb may appear almost anywhere in a sentence, and often its position affects the meaning of a sentence. For example, compare the adverb *almost* in the following pair of sentences: **Unscrew almost all of the bolts** and **Almost unscrew all of the bolts**. Compare **You positively cannot identify the source of the problem** and **You cannot positively identify the source of the problem.**

Only is a particularly troublesome adverb. Be sure to place it before the word or phrase it modifies. For example, compare the two following sentences: **Only the compare program failed to complete** means that the compare program alone of all the programs failed to complete; **The compare program failed only to complete** means that the compare program ran fine up to the point of its completion.

To stress an adverb, place it before the subject of the sentence, for example: **Clearly, the Super-10 has simplified editing**. This construction is most generally used in marketing publications.

In the last example sentence, is "has" part of a present perfect ("has simplified") or a verb (meaning "possesses') on its own? How would you read "simplified" differently in each case?

If you place an adverb at the beginning of a sentence, then set it off with a comma, for example: **Throughout, the design process is iterated**.

Read the last example without the comma. See the difference?

3. *Appositive.*   An appositive, the restatement of a word or phrase with another word or phrase, should immediately follow the noun it modifies.

Avoid using *or* as the lead word in an appositive even though you have correctly set off the appositive with commas; your reader well may not know the rule and be confused as to whether the appositive is a restatement or a choice.

Did you notice the appositive in the first sentence under "Appositives"?

4. *Relative Clauses.* Relative clauses—which are introduced by the relative pronouns *what, whose, who, whom, that,* and *which*—should immediately follow the noun that they modify.

In what order do you expect to see the above series discussed?

Don't use *what* in sentences like [The following paragraphs tell what steps to take]; instead use *the.*

Whose is used in structures like **Call the person whose duty it is to troubleshoot to the next maintenance level**.

The choice between *who* and *whom* depends on whether the relative pronoun is the subject or the object of the relative clause. For example, in **This guide is to be used by people who will install the hardware**, "who" is the subject of the verb "will install." In **The field service rep whom you assist will have run the initial error checking routines**, "whom" is the object of the verb "assist."

The difficulties arise most with *that* and *which. That* introduces restrictive relative clauses; *which* introduces nonrestrictive relative clauses. The first sentence after "4. Relative Clauses," which discusses relative clauses, is an example that contains both a restrictive and and a nonrestrictive relative—as is this sentence.

Restrictive relatives identify the noun they modify, so they are essential to the meaning of the sentence. Nonrestrictive relatives provide additional information about the noun they modify; in effect, they are nonessential and, therefore, are set off by commas.

The choice of relative or nonrelative pronoun and punctuation can change the meaning of the sentence, for example: **The CPU that recognizes its ID takes control** means that more than one CPU exists in the system. **The CPU, which recognizes its ID, takes control** means that only one CPU is under discussion.

Do you think that your reader distinguishes between *that* and *which*? Most often not, I suspect, but a technical point in grammar should be important to a technical writer. Understand the distinction before you decide its importance.

5. *Prepositional Phrases.* Prepositional phrases normally modify nouns and verbs, so they usually function as adjectives and adverbs.

   Avoid stacking more than two prepositional phrases in a row. For example, [Insert the connector of the cable with the three prongs to the three-hole plug on the left of the frame.] stacks five.

   You can replace a prepositional phrase that begins with *of* by making a noun possessive, ("the connector of the cable" becomes "the cable's connector") or by using hyphens ("cable-connector") or by turning nouns into adjectives ("prongs" becomes "pronged").

   Does the cable have the three prongs, or is it the connector? Probably the former, so the sentence could be recast to read **Insert the three-pronged cable-connector to the three-hole plug on the frame's left**. Balancing adjectives and prepositional phrases within a sentence produces a wide variety of modification. Practice a few variations.

6. *Participial Phrases.* Participial phrases modify nouns or pronouns. If they do not clearly and logically refer to the correct noun or pronoun, then a dangling or misplaced modifier results. A dangling modifier's referent is missing; a misplaced modifier appears to modify the wrong word or phrase.

   In the sentence [After replacing the disk, the system must be reinitialized], the phrase "replacing the disk" precedes the noun "system," which is not its logical referent, so the phrase dangles. To correct the sentence, provide the referent, as in **After replacing the disk, reinitialize the system**. (The logical referent is unstated second person, imperative.)

   In the sentence [Kept clean, you can use the floppy indefinitely,] the phrase "kept clean" seems to modify the pronoun "you," so the phrase is misplaced. To correct the sentence, place the participial phrase next to the word that it modifies, for example: **Kept clean, the floppy can be used indefinitely**.

Some grammarians are crusading for an end to harangues on dangling and misplaced modifiers, saying that the sentence meaning is seldom obscured by them. What do you think?

11.3.2 Paragraphs

A paragraph is a group of sentences that should form a logical structure in that each paragraph deals with a single topic, which is generally mentioned in the first sentence of the paragraph, the "topic sentence."

To test whether a paragraph is a logical unit, try giving each paragraph a temporary title; then include the words you came up with for the title in the topic sentence.

Historically, paragraphs have been classified as narrative, descriptive, supportive, comparative, and climactic.

In what order do you expect to see paragraph classes discussed?

NARRATIVE PARAGRAPHS rarely are required in technical publication writing; they are quite common in technical articles that chronicle an experiment though.

DESCRIPTIVE PARAGRAPHS, which tell how an object looks or acts, and SUPPORTIVE PARAGRAPHS, which include a general statement of the topic followed by details, are the most commonly used.

COMPARATIVE PARAGRAPHS, whether pro, con, or both, discuss the advantages and disadvantages or normal and abnormal conditions of some topic.

CLIMACTIC PARAGRAPHS are the opposite of supportive paragraphs. Like periodic sentences, which save the independent clause till the end, climactic paragraphs save the topic sentence to the end of the paragraph. They are better left to mysteries than used in technical publications.

If you introduce two subjects at the beginning of a paragraph, then either discuss neither or both of the subjects in that same paragraph. Often—and incorrectly—a writer will introduce two subjects, discuss one, and start a new paragraph for the second. If the subjects are short, then discuss both in one paragraph along with or following the introduction; if they are long, then start a new paragraph for each subject.

A paragraph of technical prose should not exceed ten lines. One sentence paragraphs are acceptable, especially between long paragraphs.

11.3.3 Transitions

Transitions are references to what is going to be said or what has been said. A book, in fact, can be considered a series of transitions from introduction to conclusion. The transitions should follow naturally from content—but they can be helped along by words and phrases, clauses and sentences, and paragraphs.

What sequence and sets do you expect the discussion of transitions to follow from the structure of previous introductory series?

Some WORDS and PHRASES are inherently transitional:

> *If* and *insofar as* signal conditions.
> *Because* and *if* signal cause.
> *As a result, consequently, hence, so, then, therefore,* and *thus* signal results.
> *In order to* and *so that* signal purpose.
> *a, b, c, . . .* ; *finally, finished; first, second, third; next, now, then*; and *1, 2, 3, . . .* signal series.
> *During the time* and *while* signal synchronicity.
> *Except, except for,* and *except that* signal exceptions.

You knew that already, right? Why? You got the signal.

In other words and *that is* signal explanations.

As an illustration, for example, for instance, and *specifically* signal examples.

Correspondingly, equally, in comparison, in the same way, likewise, and *similarly* signal similarity.

Alternately, although, but, conversely, however, in contrast, in spite of, instead, rather, and *unlike* signal a statement of opposition with what has preceded.

Again, also, and, furthermore, in addition, moreover, too, and *what is more* signal additional, often reinforcing, information.

Altogether, in short, overall, then, and *to summarize* signal a summary of preceding information.

Following signals lists, figures, tables, and topics.

In addition to containing the words and phrases above, CLAUSES and SENTENCES can refer to previous text by pronouns and repetition of nouns.

PARAGRAPHS that form a step in a procedure, show a cause and/or effect, note an exception, or provide a summary are the transitions internal to each section of the book. One-sentence paragraphs can effectively serve as introductions or conclusions to longer paragraphs.

11.3.3.1 Introductions

Technical publications, the topics that they are about, and the topics within them need to be introduced.

The preface introduces the publication.

The introduction to the topic of a publication, a level 1 section usually titled "Overview" or "Introduction" is essential, but it should not generally exceed five pages.

Of course the first sentence in an introductory section will contain the name of the product you are writing about. Even though you have probably already provided the acronym in the preface, place the acronym (in parens) after the first mention of the product's spelled-out name again. The purpose of the book and the intended audience, also both stated in the preface, should be repeated in the first paragraph of the introduction.

If the book is to be a marketing publication, then add a little hype, but try to avoid calling your product whatever the prevailing buzzword, like "state-of-the-art" or "friendly," is.

If the book is to be a set of procedures, then introduce their overall purpose and the general sequence of high level procedures that will follow in order to achieve that purpose.

If the book is to be a set of hardware or software capabilities, then introduce the set of high level functions that will be discussed in following sections.

When you introduce a series of topics in the publication, be sure that the introductory list or series names the topics in the same order as they are discussed.

When you introduce a series in a sentence, state the classification first and then the members of the class, for example: **The entry includes the record ID, the record length, and the member name**, not [The record ID, the record length, and the member name are included in the entry].

You might note that the list of transitional words and phrases in Section 11.3.3 was not presented in accordance with the previous rule. Can you see the reason for their presentation? I hoped you would anticipate the classifications, as that is what the transitional words signal.

11.3.3.2 Conclusions

Conclusions are generally not part of technical publications. Consider using a sentence or a short paragraph at the end of procedures, for example: **Now that the diagnostics have run successfully, you have completed installation of the Micromecca Super-10** or **When you receive a condition code of 0 in response to your query, the job has run to completion**.

11.3.4 Parallelism

Parallelism, the use of like grammatical structures to express items that are at an equal level, is the foundation and superstructure of technical writing—no matter the topic.

Lincoln's Gettysburg address, Martin Luther King's "Letter from Birmingham Jail," and John F. Kennedy's "Ask not what your country can do for you, but ask what you can do for your country" are paradigms of parallelism.

Parallel structure helps readers to anticipate, understand, and remember material even though they may be unconscious of the pattern. Parallelism is good rhetoric and good mnemonic.

Once items have been defined as parallel—whether they be subassemblies composing an assembly, routines composing a program, or fields composing a record—their presentation should mirror that parallel value in:

Headers
Point of view
Words
Phrases
Clauses
Conjunctions
Patterns

11.3.4.1 Parallel Headers

Keep the headers of each level in parallel structures, and keep their content at a parallel level of detail. The following two examples illustrate parallel headers:

Example 1:

SECTION 2—TEXT COMMANDS
    2.1—EDIT COMMANDS
        2.1.1 *Create*
        2.1.2 *Change*
        2.1.3 *Move*
           ·
           ·
           ·

2.2—FILE COMMANDS
    2.2.1 *Copy*
    2.2.2 *Delete*
      .
      .
      .

SECTION 3—GRAPHICS COMMANDS
  3.1—GRAPHS
    3.1.1 *Define*
    3.1.2 *Draw*
      .
      .
      .

  3.2—CHARTS
      .
      .
      .

The above example uses noun phrases as level 1 and level 2 headers and command names (imperative verbs) as level 3 headers.

Example 2:

SECTION 2—UNPACKING
    2.1—OPENING THE CRATES
        2.1.1 *Remove the Nails*
        2.1.2 *Disassemble the Crate*
        2.1.3 *Check the Contents*
    2.2—LAYING OUT THE CABLES
      .
      .
      .

SECTION 3—CABLING

The above example uses gerunds as level 1 headers, gerundial phrases as level 2 headers, and instructions (imperative sentences with header notation) as level 3 headers.

11.3.4.2 Parallel Point of View

Keep the point of view parallel.

For example: **Loosen the four bolts, slide them out of their brackets, and then remove the L-2 assembly** is second person point of view. [Loosen the four bolts, slide them out of their brackets, and then the assembly must be removed] incorrectly switches from second to third person (from imperative to declarative) and from active to passive voice.

11.3.4.3 Parallel Words

Keep parallel adjectives in parallel structure.

For example: **The output reports are accurate and readable**, not [accurate and easy to read], shows parallel adjectives joined by "and."

11.3.4.4 Parallel Phrases

Keep phrases in a series in parallel structure. A common error in parallel structure occurs because of mixed verbal phrases. The following two examples illustrate parallel phrases.

Example 1

> **One chemical bath is necessary for etching the polyimide and another for etching the metal** is grammatically correct because it contains a parallel pair of gerundial phrases (which are part of a prepositional phrase).

What part?

[One chemical bath is necessary to etch the polyimide and another for etching metal] incorrectly switches. It would be equally correct to place both instances of "etch" in infinitive form.

Example 2

> **The major steps in repairing solder bumps are to remove the excess solder, clean the area, and apply new solder** is grammatically correct because it contains a parallel series

of infinitive phrases (even though the *to* marker is not repeated).

[The major steps in repairing solder bumps are to remove the solder, clean the area, and applying new solder] incorrectly switches. It would be equally correct to place all three phrases in gerundial form.

To strengthen a parallel structure, you can repeat an article, a pronoun, a preposition, or the *to* before the infinitive, for example: **. . . to remove the solder, to clean the area, and to apply new solder**.

11.3.4.5 Parallel Clauses

Keep clauses in a series in parallel structure.

A common error in parallel construction occurs because of mixed clauses and phrases. For example: **The CSECTS that make up the subroutines, the subroutines that make up the modules, and the modules that make up the application are all available to other programs in the system** contains a parallel series of relative clauses. [The CSECTS making up subroutines, the subroutines that make up the modules . . .] incorrectly mixes a participial phrase with a relative clause.

11.3.4.6 Parallel Conjunctions

Keep independent clauses that precede and follow coordinating conjunctions *and*, *or*, and *but* parallel. For example: **The power is automatically shut down, and the alarm is automatically activated** is correct, but [The power is automatically shut down, and the alarm activates automatically] is incorrect because of two inconsistencies: voice and adverb positions.

Keep words, phrases, and clauses that are connected by a pair of correlative conjunctions in the same grammatical form. For example: **Each switch is either on or off, never undetermined** is correct, but [Each switch is either in on position or off, never undetermined.] is incorrect.

Do you see where meaning and grammar meet (that "on" and "off" are opposites, therefore parallel)?

11.3.4.7 Parallel Patterns

Keeping a pattern of notation and format like in like sections of a publication not only makes your reader's work easier, but also makes your work easier. Once you have established the pattern, the boilerplate, just plug in the information: instant parallelism.

The following is a boilerplate for presenting a group of commands:

> COMMAND NAME (Preferably at a header level contained in table of contents)
> Definition: (a phrase)
>   Syntax: (using notation introduced previously)
>   where: (explanation of operands and parameters)

The following example follows the boilerplate:

TIMESTAMP
Definition: prints month, day, and year on each page.
  Syntax: Timestamp [filename] [TOP I BOT]
where:
  filename - name of the file to be dated if that file is not currently open.
      TOP I BOT - whether the date is to appear on the top or bottom of the page. "TOP" is the default.

Of course the pattern can be expanded to include characteristics common to several commands or to comment on particular groups of commands, for example:

> Function Keys: (numbers or names or both)
>   Required: ("yes" or "no")
>   Comments: (peculiarities or limitations)

Error codes and messages also fall into a pattern. The following is a boilerplate for presenting error codes:

Code: (exactly as it appears)
Cause: (all the possibilities)
Action: (user's response to all the possibilities)

The following example follows the boilerplate:

Code: F9
Cause: Readback from test pattern incorrect.
Action: Replace PROM9.

If there is additional interaction between the user and the system, then add:

Response: (System's possible responses to user action)
   Action: (user's possible responses to system action)

for as many interactions as may be necessary, for example:

Message: ID NOT RECOGNIZED
   Cause: 1) You have entered the ID incorrectly, or
            2) You have not reactivated your ID within the allotted time.
   Action: Reenter ID.
Response 1: PASSWORD?
   Action 1: Enter the password to complete logon.
Response 2: ID NOT RECOGNIZED.
   Action 2: Reactivate the ID by entering "LOGCHANGE". (See LOGCHANGE command to reactivate your ID.)

Sections 3.5.2 and 3.5.3 exemplify some slightly different command and error message patterns. Choose a set that you think works best, design a pattern that suits the subject you're writing about, and use the pattern consistently.

12

DICTION

Diction is all the words in the publication.

The inhabitants of the electronics world are more than aware of diction. They create words all the time: company names, product names, and all the arcane vocabularies for countless disciplines. Claude Shannon, an information theorist, gets credit for *bit*; *BUNCH* (for Burroughs, Univac, NCR, CDC, and Honeywell) is attributed to Ulric Weil, an investment analyst. Sorcim Corporation is *micros* spelled backwards; Soroc Technology, Inc. is an anagram of *Coors*; Atari is a term in the Oriental game *Go*. Xerox, one of the first companies to use an *X*, is named for the Greek word *zero*, meaning "dry."

English contains the largest vocabulary of all the world's languages; it is also the language of technology. You will be choosing some of the technical terms and most of the other words in your book.

12.1 CONSISTENCY

Always use the same term or word to refer to the same concept or operation or object. For example, don't use *access*, *fetch*, and *read*; or *turn*, *rotate*, and *close*; or *memory*, *storage*, and *cache* interchangeably. Pick one and stick to it, for example: *memory* to mean "internal main memory" (what used to be called "core"); *storage* to mean "on disk and/or tape," and *cache* to mean "the buffer that holds information between memory and the logical CPU."

12.2 TERMINOLOGY

Terminology is the words used within a specific discipline, as the terms in Section 11.1, which apply to English grammar.

Terms used within someone else's discipline are usually referred to as jargon. If the jargon is well understood in a discipline, then there is nothing wrong with using it. In fact jargon is well worth using where it gains more in better communication than it costs to clarify its meaning; ditto for acronyms, as when the reader recognizes **CPU** whether it is used to stand for **c**entral **p**rocessing **u**nit, **c**entral **p**rocessor **u**nit, or **c**ontrol **p**rocessing **u**nit. (A CPU is like a rose.)

Don't be intimidated by readers—or writers—who snicker at *interfacing* or *keyboarding*.

Interfacing deserves to be snickered at only when it means "talking with," but if it is used to mean that two people share a common boundary, that their jobs are dependent on inputs or outputs from each other, then it is appropriate.

Keyboarding has a specific meaning, "entering information at a computer or terminal." If *typing* is the genus, then *keyboarding* is the species. Once typing on a typewriter disappears, the word *keyboarding* may well replace *typing* or vice versa.

Don't necessarily invent new verbs from old nouns or technical terms, but many words already function as both nouns and verbs—for instance, *program*, *bolt*, and *contact*—and more are being accepted all the time, for instance, *torque*.

Don't blush at using an acronym as a verb. How better to say "IPL the system"?

Never trust a technical term. Because we have come from so many different electronics companies, ask more than one person what a term means before using it. If you get varying answers, then explicitly define the term the first time you use it, add it to the glossary, and use it consistently. Chapter 13 discusses definitions.

If you get the chance to name a software or hardware component, then assign a unique term, the most descriptive term possible. Base your choice on the item's attributes, including function, location, or sequence, for examples: **Operand2 Input Register**, **A-14 Card**, **Chip Test #5**. Add the term to the glossary and use it consistently.

If you attach a prefix or suffix to an existing term in order to coin a new term, the affix should have a sound etymological basis. For example, "Epicode" should mean code above or attached to existing code. Add the new term to the glossary and use it consistently.

12.3 PRECISION

Mark Twain said, "The difference between the right word and the almost right word is the difference between lightning and the lightning bug." Precision in diction means choosing the right word.

If you are unsure of a word, then use a dictionary or thesaurus.

Avoid all-encompassing words like *all*, *always*, *every*, *never*, unless you are sure they are accurate. Because of the please-don't-eat-the-daisies rule (you'll seldom think of all the possibilities), it's well nigh impossible to be sure.

The following pairs and triplets can be confused because of their closeness in spelling or meaning or both. For want of a better heuristic, they are arranged according to the spelling of the first word in the set and explained by part of speech and/or meaning and/or grammatic function.

Notice the notation in the following list. Is it a coherent pattern?

accept = verb meaning "to consent to"
except = preposition meaning "excluding"

adapt = verb meaning "to adjust to something new"

adept = adjective meaning "highly skilled"

adopt = verb meaning "to take as one's own"

affect = a verb meaning "to influence"

effect = a verb meaning "to bring about" and a noun meaning "a result"

affinity = a noun meaning "attraction"

aptitude = a noun meaning "talent"

all ready = an adjective meaning "prepared"

already = an adverb meaning "previously"

all right = each and every one correct

[alright] = not a word—yet

alternate = a verb meaning "to occur in successive turns"

alternative = a noun meaning "the choice between two mutually exclusive selections"

all together = a group acting together or in one place

altogether = entirely

although = even though

while = at the same time

among = a preposition for distinction among three or more elements

between = a preposition for distinction between two elements

ante- = a prefix meaning "before" or "in front of"

anti- = a prefix meaning "against" or "opposed to"

any = a choice among three or more elements

either = a choice between two elements

as = a subordinating conjunction used before a clause

like = a preposition used before a word or phrase

aspect = a quality of a larger subject

phase = a stage of transition or development

attribute = a noun meaning "a quality belonging to something" (accent on the first syllable)

attribute = to point to a cause (accent on the second syllable)

contribute = to give

augment = to increase or magnify in size, degree, or effect

supplement = to augment in order to compensate for a deficiency

capacity = measure of volumes and amounts

capability = measure of what can be done

cite = to acknowledge or to refer to

site = the place where something is located

sight = the ability or action of seeing

compare = to point out similarities and differences

contrast = to point out differences

complement = verb or noun meaning "something added in order to complete"

compliment = verb or noun meaning "praise"

connote = to imply

denote = to state directly

continual = recurring with interruptions

continuous = occurring without interruption

contiguous = continuous in space

credible = believable

creditable = worthy of credit

criteria = more than one standard for judging or testing
criterion = one standard for judging or testing

defective = faulty
deficient = lacking a necessary ingredient

diagnosis = an analysis of the nature of something
prognosis = a forecast

different from = an introduction to a noun or noun phrase
different than = an introduction to a clause

discreet = showing careful or prudent behavior
discrete = separate, distinct, individual

e.g. = for example (spell it out in text)
i.e. = that is (spell it out in text)

electric = pertaining to wires and cables
electronic = pertaining to tubes and transistors

emulation = one machine copying another machine (the underpinnings of compatibility)
simulation = a model corresponding to a different kind of a model of reality (for example, mathematical and graphics models of the same subject)

explicit = directly stated, denoted
implicit = implied, connoted

farther = distance in space
further = distance in time, direction, or abstraction

feasible = can be done
possible = can happen

fewer = an adjective describing count nouns
less = an adjective describing mass nouns or an adverb

flammable = capable of being set on fire
inflammable = capable of being set on fire
nonflammable = incapable of being set on fire

imply = to suggest
infer = to draw a conclusion

in = within
into = movement from the outside to the inside

insoluble = incapable of being dissolved
insolvable = incapable of being solved
unsolvable = incapable of being solved

its = belonging to *it*
it's = contraction of *it is*
[its'] = NOT a word

liable = legally subject to; responsible for
likely = probable

loose = not fastened
lose = to be deprived of

may be = verb indicating possibility
maybe = adverb meaning "perhaps"

medium = one means of communication or storage
media = two or more means of communication or storage

method = a way in which to do something
methodology = the study of methods or an overall heuristic
           for a combination of methods

mid = adjective (for example, "mid board")
mid- = adjective-noun unit used as adjective (for example,
       "mid-level signal")
mid- = prefix (for example, "midpoint")

notable = worthy of notice
noticeable = readily observed

observance = following a rule
observation = notice of something

on = supported by, attached to
on to = direction in space or time
onto = movement to a position

people = a group
persons = individuals
personal = an adjective meaning "pertaining to an
           individual"
personnel = a collective noun meaning "a group of people
            working at a common job or company"

phenomenon = one observable thing or event
phenomena = two or more observable things or events

practicable = possible, feasible
practical = useful

preceding = coming ahead of
proceeding = continuing

prescribe = dictate, suggest
proscribe = prohibit

principal = an adjective meaning "primary"
principal = a noun meaning "a person in authority" or (for
           tech writers working at startups) "a sum or
           money," as an investment
principle = a noun meaning "a rule"

proved = past and past participle verb form of "to prove"
proven = adjective form of "to prove"

pseudo = false or counterfeit
quasi = somewhat
semi = half

some time = an amount of time
sometime = an unspecified time
sometimes = on occasional occurrences

stationary = not moving
stationery = writing paper and envelopes

there = an expletive or adverb
their = a possessive pronoun
they're = a contraction of "they are"

to = a preposition or mark of the infinitive
too = an adverb meaning "also"
two = the numeric 2

12.4 GENDER

Avoiding sexism in writing means recognizing that language has the power of serious discriminatory implications. A word like 'field service rep" has by convention taken a masculine pronoun, but now that men and women both are reps—and programmers, and engineers, and technicians—language should change to reflect the change in society.

Using "he/she" and "him/her" is awkward and distracting; the "s/he" construction is clever, but then what to do with "him/her"? Repeating the antecedent noun is acceptable, but pretty ponderous.

The commonsense ways to eliminate sexism in writing are to use the second person as much as possible—a good idea in technical writing in general—and to avoid singular personal references, especially since most grammarians still say that *they*, *them*, *their* are not acceptable after a singular antecedent.

In some cases, using new diction is jarring, but language will change for the better when "people hours" or "staff hours" or "worker hours" sounds as ordinary as "man hours" does now.

It is a credit to the electronics professions that there never have been programmesses or enginettes.

12.5 ECONOMY

Be economical, concise, but don't sacrifice meaning to economy or make your writing abrupt. For example, if "Close cover" seems cryptic or impolite, then write **Close the cover**, or **Close the cover firmly**, or **Close the cover one full turn**, but [Utilize maximum care to ensure that cover is held firmly in position] is overkill.

Economy of diction is usually the easiest part of bettering your book. You can use Table 12-1 to compare to your draft. Simply change all of the occurrences of the constructions that appear in your draft and in the left column of the table to the constructions in the right column.

Table 12-1. Economy

Cross Out	Use
A small number of	a few
absolutely essential	essential
actual fact	fact
adequate enough	adequate
advance warning	warning
all of	all
arrive at a decision	decide
as to whether	whether
ask the question	ask
at the present time	now
at this point in time	now
cannot help but	cannot avoid
completely eliminated	eliminated
conclusive proof	proof
conduct an investigation	investigate

Table 12-1. Economy (CONT.'D)

Cross Out	Use
couple together	couple
despite the fact that	although
during the time that	while
existing conditions	conditions
few in number	few
for the purpose of providing	to provide
for the reason that	because
has the responsibility of/for	is resposible for
in back of	behind
in connection with	with
inflammable	flammable
in many cases	often
in most cases	usually, generally
in the event that	if
inside of	inside
involves the necessity to	necessitates, requires
irregardless	regardless
is due to the fact that	because
join together	join
make an approximation of	approximate
on account of	because
on the grounds that	by
on the part of	by
open up	open
orientate	orient
outside of	outside
preventative	preventive
rarely ever	rarely
red in color	red
rough in texture	rough
round in shape	round
seldom ever	seldom
small in size	small
take into consideration	consider
the reason is because	because
through the use of	by
totally complete	complete
undertake a study of	study
until such time as	until
utilize	use
with a view to	to
with the exception of	except
with the object of	to; in order to

12.6 AESTHETICS

Even though your publications are not likely to be read aloud, be aware of alliteration or rhyme. For example, avoid constructions like [processors principally provide] and [function in connection with protection inspections] and [effect with respect to neglect].

13
DEFINITION

A definition may be a word or group of words, or it may make up the bulk of a document, as in a product specification. Don't defeat the purpose of your job, which is supposed to be making someone else's job easier, by using difficult diction, as in Samuel Johnson's infamous definition: "A network is anything reticulated or decussated, at equal distances, with interstices between the intersections."

Though examples should be used whenever possible to expand definitions, they do not take the place of the definition.

No matter how well you identify your audience, they will consist of a population varying from novice to expert. What definitions do they need? How do you gauge what to assume and what to define?

Use your knowledge of the subject as a gauge. Because tech writers are always learning, you will be familiar with some of the subjects you are writing about and unfamiliar with others. If you are very familiar with the subject, then you are apt to define too little; if you are totally (it happens) or somewhat unfamiliar with the subject, then you are apt to define too much. Adjust accordingly, but lean toward defining too much.

13.1 WORDS

Words are labels for ideas and objects. Some of the most common words are recognized only according to context. For example, if someone at work asks, "Where's the bus?"; then you probably look at a block diagram to find a transmission symbol. If someone at the Greyhound Station asks the same question, then you look for something very different.

Though the audiences you're writing for won't be quite that varied, you'll be judging commonality in order to define words according to the category they best fit:

1. *Common words used commonly* don't need definition. They are the ordinary words that most people share, so use them whenever possible. Even though some of your readers will hold advanced degrees from ivy-covered walls east or west, you can't expect the same of all of them. "Ameliorate" is a grand word, but "improve" is sure to be understood.

2. *Uncommon words used in common* aren't too difficult to define. They are most often terms naming parts of the product you are writing about and can be defined generically: Whether named *VM*, or *UNIX Operating System*, or *Executive Control Program*, an operating system by any other name performs most of the same functions. Keep a running glossary of these terms and their acronyms.

3. *Some common words used uncommonly* must be defined; otherwise how is your audience to know?

   First try to convince the engineer or programmer that it is unfair to burden the users who think they understand the word with having to play Alice in Wonderland. For example, if the programmer has used *initialize* to mean "reset," then your reader is likely to think the operation happens only once when, in truth, it may occur at each machine cycle.

   If the word is already hardcoded in some way (maybe through marketing documentation), then define it more than one time for those readers who skip the page when you first defined it.

Did you recognize the word "hardcoded" in the previous sentence? If so, then you'll recognize jargon. What does its use say about how I gauge my audience for this book?

4. *Other common words used uncommonly* are shared by people in one discipline who define a word's particular meaning differently from what is common to the general population, for examples: **bug**, **compile**, **convert**, **disk**, **mask**, **memory**, **read**, **write**, and **terminal**. As electronics continue to proliferate, many of these words are reaching category 1 status, common words used commonly.

   Is "electronics" a mass, count, or collective noun in the above sentence? Does the verb agree with your choice?

   Working with the technologists, you are becoming more familiar with new words every day, but be aware that different disciplines within the industry use the same word to mean different things. A *device* might mean a transistor to a circuit designer and a disk or tape drive to a systems programmer. Generally the term is clear in context, but if you are describing a full system from technology through operations, do you want to use *device* to mean both of the above?

5. *Uncommon words used uncommonly* means you are learning something new—the best kind of assignment. Keep a running glossary of these words, and as you learn more about the subject, you will know better which ones need to be defined for your readers.

13.2 WAYS

You can choose from several methods of definition: analysis, cause and effect, comparison and contrast, negation, and location. Try to choose the way in which the definition will have the most meaning to the audience. Granted, there is no "true meaning," but, practically, meaning is what the reader thinks something means.

By now you are probably expecting that the following sections will be titled and sequenced to match the series in the last paragraph. Your series should provide and fulfill the same expectation.

13.2.1   Analysis

Analysis defines by distinguishing and separating ("decomposing" as the current jargon has it) a physical object into parts, a function into subfunctions, a procedure into steps.

Physical objects must be defined in physical terms, for examples: **The connector is a three-foot long coax cable, having a plug on either end** and **The mainframe is 5′ high × 6′ wide × 29¾″ deep**.

Functions must be defined in terms of operations and relationships, for examples: **The adder adds two 32-bit numbers from the operand buffers and stores the results in the results buffer** and **The laser beam writes depressions or bumps on an optical disk**.

Procedures must be defined in terms of their results and the steps to get there, for examples: **In order to install the update tape, you will unpack it, load it, and initialize it** and **To analyze the error, perform the following set of diagnostics**.

13.2.2   Cause and Effect

Most operations and commands are defined in terms of their effects, for examples: **INIT sets all registers to 0** and **The "File" command removes the data from the screen and stores it on disk**.

If you are writing about a conditional cause and effect, then use *If . . . , then . . .* constructions, for examples: **If you choose the "print" option, then the following screen will be displayed** and **If the condition code is more than 1, then either the program or the data is in error**.

13.2.3 Comparison and Contrast

It is often helpful to define the unfamiliar by comparing it to the familiar—likenesses and/or differences—especially if the term you are defining is a relatively new species of a known genus, for example: **The biadic processor is essentially two uniprocessors under control of one operating system; it functions like a multi-processor.**

Comparative statements are analogies, which often contain the word *like* ("similes" in literary jargon). For example, you might define a *read* by **The read operation is like reinserting a page in your typewriter in order to make additions or corrections**.

Contrastive statements are opposites ("antonyms" in grammar jargon). For example, you might define a *write* as **The write operation stores the file on disk, opposite to the read operation, which displays the file on your screen**.

To what audience would the two previous examples have been written? If you said "Probably to a very new user of a wordprocessing system who is used to a typewriter," then I wrote it right.

13.2.4 Negation

If the word to be defined has closely related terms, or is a common term used uncommonly, or is opposite to another word, then you can tell your audience what it is not, but state what it is immediately following, for example: **A net is not necessarily a complete circuit; it may be only one wire connecting an output terminal to an input terminal.**

13.2.5 Location

Physical objects may be defined by location as well as by shape and size, especially in maintenance manuals, for example: **The P102 cable is connected at J-10 and J-14**.

13.3 FORMS

You may define terms explicitly or implicitly.

13.3.1 Explicit

Explicit definitions are formal statements of what a word or phrase means at the appropriate level. For example, **The CPU is contained on three boards** defines "CPU' in physical terms; **The CPU processess System/370 instructions** defines "CPU" in functional terms.

An explicit definition uses words like *is a*, *means*, *results in*, and *consists of*.

Glossaries contain explicit definitions, but they are also contained in sentences and paragraphs in text.

Is there any ambiguity in the pronoun in the previous sentence? If you think so, then you could repeat "explicit definitions."

13.3.2 Implicit

Implicit definitions are asides; they provide a familiar word, phrase, or clause for the unfamiliar term just used in text. They are especially useful for those words you are unsure about whether to define. If some of your readers don't know the term, then it will be helpful. If other readers already know the term, they merely have to read an extra word or few.

Appositives and nonrestrictive relative clauses are the common forms of implicit definition.

NOTE: Remember to use commas, parens, or dashes to set off appositives and nonrestrictive relatives from the rest of the sentence.

13.3.2.1 Appositives

Appositives, words or phrases that immediately follow a term and restate it, are the most informal of definitions.

For examples: **Each node (terminal) carries a unique alphanumeric identifier** and **A reticle—a particular configuration of 400 array sites—is dedicated either to logic or to RAM**, and the first sentence of this section contain examples of appositives used as implicit definition.

Avoid using *or* as the lead word in an appositive. Though you know commas set off appositives, many of your readers won't and might mistakenly think the sentence implies choice.

If you've read the section on modification, then you have already encountered this caveat. Do you think it's worth stating twice? Why? The "If" at the beginning of this consideration should give you a clue.

13.3.2.2 Nonrestrictive Relatives

Relative clauses, which are dependent clauses that immediately follow a noun, are clearer than appositives as definitions—but wordier.

For examples: **Each node, which is a terminal, carries a unique alphanumeric identifier** and **A reticle, which is a particular configuration of 400 array sites, is dedicated either to logic or RAM**, and the first sentence of this section contain examples of nonrestrictive relative clauses used as implicit definitions.

NOTE: Restrictive relative clauses should not be used to define, but to identify, items. For example, "A terminal that is a node" means that there exists a terminal that is not a node. If that is the case, then the construction is correct, but it is not a definition.

14
DESCRIPTION

Description is a type of exposition, which is any writing that attempts to explain a subject accurately and completely.

Description deals specifically with details, but before you supply them, tell the reader about the overall system. The length of the introductory exposition depends on your audience, but in any case, postpone detail until the readers need to use it, to save them from searching back and /or reading dreary, "as mentioned earlier" phrases.

This chapter discusses the descriptions most common to technical writing:

Physical description: HOW the parts of the system fit together

Functional description: HOW the system works

Procedural description: HOW TO make the system work

14.1 PHYSICAL

A physical description discusses an item's dimensions and interconnections and often accompanies a procedural description in installation and maintenance manuals.

Hierarchical classifications, the systematic arrangement of related parts in a larger unit; and spatial partitioning, the division of a unit into its parts, are the essential structures of physical description.

NOTE: A unit must contain at least two parts, and each part may belong to only one unit.

Begin a physical description with a short explanation of the use of the product, for example: **The Super-10 is a complete workstation for data and graphics**. Provide a photograph or illustration of the product as part of the introduction.

Proceed from the general to the specific, starting with the fewest physical units, for example: **The Super-10 is composed of three major units: the Control and Processing Unit (CPU), the Monitor Terminal Unit (MTU), and the Hard Disk Unit (HDU).**

Discuss each physical unit in turn—in the same order that you introduced it—and, again, partition the unit into the fewest physical parts, for example:

> **The CPU is composed of six assemblies:**
> **Slot Frame (SF)**
> **Control Card (CC)**
> **Execution Card (EC)**
> **Memory Card (MC)**
> **Bus Card (BC)**
> **I/O Card (IOC)**

NOTE: If there are more than three parts composing a physical unit, then put them in a list.

Discuss each part in turn—in the same order that you introduced it—and, again, partition the unit until you reach the lowest level in the hierarchy.

In order to describe each part in detail, provide its function, its size, its shape, and how it physically connects to other parts, for example: **The CC monitors and controls the flow of signals internal to the CPU. It occupies the third card slot in the slot frame.**

Use hardware illustrations or photographs to show shapes,

sizes, and relationships in the mechanism. Make sure the dimensions and the callouts in the figures match the descriptions in text.

14.2 FUNCTIONAL

A functional description discusses an item's operation, in terms of internal components, communication among them, and communication to the item's outside world. Functional descriptions form the basis for theory of operations publications and often accompany a procedural description in troubleshooting publications.

As in physical description, hierarchical classification and partitioning compose the overall structure of functional description; however, the operation of the units and their parts is the essence of functional description: the inputs, outputs, and what happens between.

Begin a functional description with a short explanation of the use of the product, for example: **The Hard Disk Unit (HDU) stores serial digital data on a 1110 cylinder nonremovable disk stack.** Provide a system block diagram illustrating the major functional areas as part of the introduction.

Proceed from the general to the specific, starting with the fewest functional units, for example: **The HDU is composed of the following major functional units: Logic Circuitry, Servo Mechanism, Power Control.**

Discuss each functional unit in turn—in the same order that you introduced it—and, again, partition the unit into the fewest functional parts, for example:

> **The Logic Circuitry includes the following functions:**
> **System Controller**
> **Input/Output Controller**
> **Bus Driver**
> **Buffer Memory**
> **Clock**

NOTE: If there are more than three items composing a functional unit, then put them in a list.

Discuss each function in turn—in the same order that you introduced it—and, again, partition the functions useful to your readers, for example:

The System Controller performs the following functions:
Instruction fetch
Instruction decode
Operand fetch
Arithmetic/logic computation
Results broadcast

Discuss each function in turn—in the same order that you introduced it—and, again, partition the functions until you reach the lowest level in the hierarchy.

In order to describe each function in detail, first describe it as a successful sequence of events; then describe fault sequences, machine malfunctions, or program errors. For example, a store operation could be described under the subfunctions of "Address Selection," "Key Matching," "Parity Checking," and "Data Transfer," followed by sections titled "Key Mismatch" and "Parity Errors."

Electronic components are active, they "amplify," "buffer," "compare," "differentiate," "accept," and "output"; so choose active verbs and predominantly active voice in functional descriptions.

Use block diagrams and flow charts to show how functions relate and operate. Make sure the figures match the description in text. Check inputs and outputs especially.

14.3 PROCEDURAL

If most technical writing is description, then most description in technical writing is procedural. Procedural description is the essence of publications about the processes of installing, operating, maintaining, and repairing equipment and systems.

NOTE: Get hands-on experience for all procedures. If that is not possible, then watch someone else perform them.

Unlike physical and functional description, the structure of procedures is sequential. The order of the steps, of course, is chronological.

Unlike the third person used for physical and functional description, second person is the voice of procedural description because instead of talking about something, now you are talking to someone, someone who needs to solder electrical connections on a printed circuit board, someone who needs to learn how to use a text processing system, someone who needs to install, maintain, or troubleshoot a machine or a program.

Because of audience variation—they may be new on the job or seasoned veterans—two specialized forms have been developed for procedural descriptions: tutorial and cookbook. A tutorial describes every move, which is good for real beginners but frustrating and tedious for intermediate and expert users. A cookbook describes the uses of the moves and allows the reader to combine them.

Compare the following examples:

Tutorial approach:
To open the top cover:
1. **Insert a small screwdriver into each of the two small openings in the rear top corners of the cover far enough to release the spring latch.**
2. **Pull up slightly on each corner to release.**
3. **Raise the cover until the support rod inside the cover is visible.**
4. **Place the support rod in the long slender slot at the left top of the frame.**

Cookbook approach:
1. **Open the top cover.**

Whether you use tutorial or cookbook format, many of the considerations are the same.

When you write a procedural book, begin with a short statement of the reason for the entire procedure, for example: **You will be**

performing the steps to install and bring up your Micromecca Super-10.

If there are any special dangers, tools, preparations, system configurations, or environmental requirements involved in performing the procedures, state them at the beginning of the publication and again at the beginning of each section that contains the procedure, for examples:

> **In addition to the tools provided, you will need a medium-sized screwdriver.**
>
> **Allow at least four square feet of space for placement.**
>
> **You will be making electrical connections, so pay attention to each warning and caution.**

As with physical and functional descriptions, procedural descriptions proceed from the general to the specific, but much more quickly. Begin with the fewest principal procedures, for example: **Installation procedures consist of unpacking the system, connecting the cables, and powering on the system.**

Discuss each item in turn—in the same order that you introduced it—and, again, partition the item into the major procedures, for example:

> **Unpacking the system consists of:**
> **Uncrating the units,**
> **Checking the contents of the crate,**
> **Placing the units, and**
> **Laying out the cables.**

NOTE: If there are more than three sets of steps composing a procedure, then put them in a list.

Discuss each item in turn—in the same order that you introduced it—and this time describe the procedure in steps.

NOTE: Don't forget that any necessary warning or caution— clearly marked—precedes each procedure or each step in a procedure requiring one.

Steps in procedures are precise actions, so choose active, precise verbs; prefer **solder**, **weld**, **plug**, and **screw** to "attach" and "connect."

Be as concise as possible, without being abrupt. The choice to use *the* or not to use *the* in procedures is still being argued. *The* doesn't take much space; it gives your reader pause between nouns; and it can disambiguate meaning. For example, contrast *Replace the spring and insert the rod* with **Replace spring and insert rod.**

"Insert" could be either a verb or an adjective. A-Ha!

Number each and only the steps that the user performs. If the system responds to the step in some way, then indicate the response at the end of the step, but don't number it. For examples:

In an installation procedure:

> **n. Remove the top of the crate.**
> **The packing envelope is taped to the top of the CPU.**
> **n + 1. Remove the Component List from the packing envelope.**

In an operating procedure:

> **n. Press the ENTER key.**
> **The system will ask for your logon ID.**
> **n + 1. Enter your logon ID.**

In a maintenance procedure:

> **n. Unhook the belt tension spring.**
> **The belt will relax.**
> **n + 1. Remove the belt.**

In a troubleshooting procedure:

> **n. Insert system diagnostics.**
> **The FE panel will display condition codes.**
> **n + 1. Check the diagnostic condition codes against the list in Appendix A.**

where: n = the step number

If a series of steps is common to more than one procedure, then refer back to long series (ten or more steps) and repeat short series.

After you have written a procedure, test it by asking someone who is as close as possible to your view of the audience to follow the steps. Plan to do some revising.

Do you see why "who" begins a restrictive relative? It's not just any old "someone," but restricted to someone specifically identified to be "as close as possible to your view of the audience."

15

STYLE

Writing style is the way in which something is written, not the substance of what is said. Good technical writing style means good physical and logical organization, correctness, clarity, completeness, and conciseness, much easier to say than do.

15.1 PRESCRIPTIONS

1. *Keep your audience in mind.* Your publications will be directed to naive and experienced readers, to technicians and Ph.D.s, to sophisticated writers of English prose and barely English speakers. Though you cannot write expressly to the temporary operator who is substituting until the regular operator returns from Squaw Valley, on crutches, if you write only for the experienced operator, then some reader who really needs all the guidance possible will find it frustratingly difficult to understand the information.

2. *Establish credibility.* A comma splice might seem to be a fussy detail, but good engineers and programmers know the importance of details to their work—it's equally important in ours.

3. *Be consistent.* Setting up patterns in structure, diction, and format helps your readers to know what to expect

151

and how to interpret it. Remember, many companies sell their products internationally, so consistency will also help translators.

4. *Make logical connections.*   Show cause and effect clearly each time that it exists—and only when it exists. Just as an OR gate will not work where the decision rests on an AND, conjunctions have the power to make or break a logical connection.

5. *Use active voice predominantly more than passive voice.* Passive voice is wordier than active voice, and language acquisition experts have shown that it's also more difficult to understand.

   Passive is often an excuse for inadequate research. Find out what does what to which, and tell all the readers all they need to know. For example, [The file is closed when the filename has been timestamped] doesn't tell whether the system or the user has to "timestamp" the file. If the system timestamps, then you might write **The system automatically timestamps the file when you enter the "CLose" command**; or if the reader timestamps, then you might write **Timestamp the file prior to entering the "CLose" command**.

6. *Avoid expletives.*   Only use expletives when their removal would cause real sentence contortions.

   Expletives are usually easily removed by recasting the sentence to begin with a verb or noun. For examples: [It is necessary to run the diagnostics] becomes **Run the diagnostics**; [There are six files on the tape] becomes **The tape contains six files**.

15.2 PROSCRIPTIONS

Don't just mimic all the technical writing that has come before; much of it is bad, and because some of it has been so frankly awful, it is easy—and fun—to parody.

Pick one from each section—I, II, III, and IV—below, and you can construct dozens of sentences that look very much like those

we've all read in technical articles and manuals, but don't adopt
their style.

I.

1. Thus, within specified program parameters,
2. From the dynamic interprocessing standpoint,
3. With respect to essential heuristics,
4. In this regard it can be maintained that
5. Based on an independent modular concept,

II.

1. a large portion of interactive communication coordinates
2. the characterization of critically cooptive criteria
3. our fully integrated support program
4. further and associated contradictory elements
5. the primary interrelationship between systems and/or subsystem configurations

III.

1. must utilize and be functionally interconnected with
2. requires considerable performance analysis and computer aided design techniques to arrive at
3. presents a valuable addition showing the positive function of
4. effects a significant implementation of
5. adds a nontrivial enhancement to

IV.

1. the system's and/or subsystem's basic configuration.
2. any communicatively programmed interactive techniques.
3. the total configurational rationale.
4. the integrity and synchronicity of data.
5. subsystem compatibility testing procedures.

Though there is some variation in diction, the above proscribed
structures are also used in papers and articles in other disciplines,

including philosophy, education, sociology, and art literary criticism.

Check the punctuation internal to the lists. Would choosing one entry from each column result in a grammatic sentence? Grammar isn't everything.

PART III: OPERATIONS

To produce a publication, you will plan, research, and write the drafts of a book, copies of which will then need to be reviewed, edited, and produced. This part discusses those operations separately and under practical circumstances.

Of course under pressure of deadlines and often having to work on more than one task and more than one book at one time, tech writers do have to write and edit information that we understand only superficially; because of the never-failing last minute changes, research is never complete; because of the last minute rush, rhetorical and copy edits are never as thorough as a good writer wants them to be.

The goal is to put out the best possible publication in the allotted time, keeping in mind that:

> The more important a piece of information is,
> the harder it is to get.

> There is usually a better way to say it,
> and often a more accurate way,
> but they are seldom the same way.

> There is always one more embarassing typo
> that every reader will see even though
> all the reviewers and editors didn't (*sic* joke).

> White space never lies.

PLANNING

The purpose of planning a publication is to ensure the readers the information they need to install, run, and fix a product when the product is ready to ship. To that end, the earlier that publications planning is integrated with product planning, the better, especially for large products that will take two years or more to develop.

16.1 PLANNERS

Programmers and engineers involved in a product's development need to plan—and commit—time to explain the product and review the drafts of the book.

Quality controllers need to plan to test the latest draft along with the latest version of the product.

Market planners need drafts to train salespeople and develop customer education before the product ships.

Publications planners must incorporate all the planners' plans in order to plan publications, keeping in mind the amount of information that needs to be monitored and the number of people who have different goals for its use.

Publication titles are generally included in product plans, but each book must be planned internally in order to schedule writing, reviewing, editing, and production time to meet the ship date—at a reasonable cost.

Though no publications planner wants to compromise the publication because of inadequate information or inadequate time for editing, in ongoing schedule negotiations with developers or marketeers, be prepared to lose gracefully. No product will be late because of a publication. There are plenty of other reasons for it to be late.

16.2 PLANS

When planning a publication or a set of publications consider your readers first.

Does your audience need to know the software interfaces in order to make the program run in their system, or do they need to know how to communicate with it at a terminal? Do they need to know how the hardware works so that they can fix it when it doesn't, or do they need to know how to push the right buttons or keys at the right time?

The following two lists name audiences and the publications they need.

1. Audiences

   Because there are many levels of sophistication and people/machine interaction the information must be tailored for the target audience. The following is a noncomprehensive list of audiences and their needs:

   Systems Programmers need complete descriptions about operating system functions, and how to modify them, presented in clear patterns.

   Applications and Maintenance Programmers need to know how to create and modify an application and how it will interact with the system.

   Technical Users, for example, computer programmers and operators, need to know what to expect from and what is required for each function as they interact with the machine to keep other people's programs running.

Maintenance Personnel may be employed by your company or by the customer. In either case, they need precise, detailed instructions and illustrations of the hardware they will be installing or fixing.

Professional Nontechnical Users need the system to be described in terms of their use for it, for instance, accounting, mechanical design, medical or geological search. They will be using the system to make their work quicker, easier (eventually anyhow), and more accurate.

Customers expect glossy, concisely written overviews of the product they are being asked to buy. Special requirements, like significant space, sensitive temperature controls, or modifications to software running in their systems need to be included.

Data Entry Personnel need nontechnical instructions in order to use the functions provided by the system.

Once you've defined the target audience, you'll have a clearer idea for the types of publications that they'll need.

Is the above list logically organized? Would it be better to state the organization explicitly (that is, generally, most to least sophisticated)? Is it more meaningful than an alpha organization? Be sure all your lists are organized in some way.

2. Publications

Because people use technical publications to perform many different tasks, the information must be tailored for the task. The following is a list of publications and their purposes:

Reference Manuals provide characteristics of a system in detail, whether it be how the circuits in the system work or how to use every instruction in the system's repertoire.

Installation Guides provide step-by-step procedures to install a product, bring it up, and make sure it works.

Operators Guides provide step-by-step procedures to teach an operator how to access a system, how to make it work, and how to recover from errors committed by the operator or the system.

Site Planning Guides provide introduction to a hardware product and describe the environment it will need, including floor space, power, cooling, and support software.

Summary Cards provide handy additions to a publication that deals with many commands or steps, for an operator's guide for a text editing system or an installation guide for a midsize machine.

System Overviews provide a prospective customer with general introductory material enthusiastically describing the concepts and general functional capabilities of the product.

Once you have determine the kind and number of publications, its time to make plans.

16.2.1 Product Plan

Your company may use Gantt, PERT, or other timeline charts for scheduling. In order to be included in these schedules at realistic junctures, you will have to understand what is really required.

What exactly is the product supposed to do? Who must know how to interact with the product? For what? By when?

Though it is important that you are interested in the internal workings of the product you are writing about, don't be preoccupied to the point that you forget its purpose—and yours—to help someone do something.

Who will be helping you? Who will be responsible for the overall project? Who will be developing what parts of the product? On whom will you be depending?

16.2.2 Publications Plan

In large, established organizations, a publications plan may be very formal, needing to be verified by the project development department, negotiated with the art department, and scheduled with production and printing departments. In small organizations, the publications plan may be seen only by the writer and the pubs manager. In start-ups, you may *be* the publications department, and the only one to see the plan.

In any case, the following information should be supplied: proposed publications, division of labor, schedule, and cost.

16.2.2.1 *Proposed Publications*

Provide a list and short description of each proposed publication, including purpose and audience. The more complex the product, the more diverse its potential audience, the more publications will be required.

People in electronics fields are wont to say that they do not care to reinvent the wheel but would rather build on proven technology: Look at technical publications that have the same purpose and audience as those you are planning. If your company has not produced a similar product yet, then they are probably using one. Look at the publications that came along with those products.

Take notes on page count and any organization and/or format elements you think are good.

16.2.2.2 *Division of Labor*

Labor on a manual must be shared to some extent by people other than writers.

To keep track of the division of labor outside of the publications department, identify:

1. The developers who will be responsible for providing information.
2. The developers and managers who will be responsible for reviewing the draft for technical content.
3. The people who will be responsible for approving the book for print.
4. Any other departments that support the publications.

NOTE: It's best to identify people by title rather than by name because the names tend to change over the course of the project.

Identify the number of writers who will be responsible for researching, writing, and editing the books. In cases of multi-volume books, books exceeding 150 pages, or an especially urgent ship date, more than one writer might be assigned to the publications for one product.

16.2.2.3 Schedule

Believe that the product is going to be shipped on time when you make out the first schedule. Gauge your end milestone date to first customer ship and work backward, including production, writing, and reviews in the schedule.

1. First estimate the amount of printing or reproduction time involved. You'll have to back up that much time from your end milestone.
2. Next estimate the time needed internal to the publications department. Publications planners often use 6.5 hours per page for the writing time gauge. The 6.5, however, includes research, illustrations, editing, and usually producing the master.

    If the material is to be written for a sophisticated audience, then 6.5 hours per page is often tight.

    If there are few or simple figures, then interfacing with illustrators takes little time, but it does take some.

    If the writer has produced similar manuals, then research and organization time can be cut significantly. But nobody likes to get stuck writing the same book over and over.

    If the writer has just switched from writing software to hardware or vice versa or has less than a year's experience, then a lot of learning will be involved with the research.

Depending on the density and difficulty of the material, the copy editor may spend an hour to get anywhere from 5 to 15 pages done.

If the pages also need formatting, then inputting the copy edit changes along with the format takes about an hour for ten pages.

To touch-up, reduce, and paste-up a figure for a master takes an average of half an hour.

NOTE: Whoever is going to be doing the production needs to be alerted to the date to expect the book in order to schedule printing and to order any materials like the front cover insert and spine tabs for binders. If the date changes, then be sure to let this person know.

NOTE: If there are any extra frills that have to be ordered or be put in order within the publication, like dividers, then try to talk the developers or marketeers out of it. It takes more time, it costs more money, and it greatly increases the chances for mix up.

3.   Don't forget to leave enough time for reviews. Before you decide when to start and how many reviews to schedule, meet the people who will be supplying the information and reviewing the drafts.

If they say, "Documentation is a pain; I don't have time for it," then surmise that they do not value your services highly and may well relegate you to the bottom of their priority list. If they say, "That's great, I'm glad you're doing it," then give them the benefit of a doubt.

If you have the time, then schedule two formal reviews of the book-in-progress: a preliminary review at about the three-quarter point in the schedule and a final review as near to the end as is feasible.

The amount of automation in the pubs department and late changes in the product either ease or expand the estimates, very seldom the former.

16.2.2.4 *Cost*

If you are in a large organization that charges one department for another's department's services, or if you are in charge of budgeting the publication, then use the standard set of cost criteria in your company.

If you're new at the job of costing, then contact the technical communications association you belong to for literature.

16.3 PROGRESS

Watch your progress tied to the progress of the developing product. Though the publications must be ready to ship with the product, product specifications, personnel, and schedules almost always change—especially if the product takes more than a year to develop.

To stay on top of how the project is progressing, attend the development staff meetings, and learn everything you can about the product.

Start a folder for the project as soon as you get the assignment. In the folder:

1. Keep a copy of the publications plan;
2. List contacts and their telephone numbers;
3. Log problems and their resolutions;
4. Store pertinent progress information—notes from meetings about schedules and requirements, all memos both sent and received; and
5. Save all signed-off review forms.

Keep this project folder separate from the notes you take on technical information.

NOTE: DO NOT THROW ANYTHING AWAY until the publication has been shipped. Keep the publications plan longer for reference the next time you have to schedule a publication, and keep the signoffs in the publication department archives, as discussed in Chapter 22.

17
RESEARCH

Before you can write about hardware or software, start understanding what you'll be writing about, for if you write without understanding, as some technical writers think they can, then you may be writing the technical version of "The Jabberwocky."

Gather all the available information about your subject from documentation and interviews.

17.1 DOCUMENTATION

Once you receive an assignment, read everything you can about the product, especially when it is something you are unfamiliar with.

If you have not written on the subject before—the most interesting kind of subject—documentation is crucial. Collect publications similar to the one you'll be writing, read through them, and editorialize so that you can write it better.

If you have written about a similar product, then you might be able to use a lot of the ideas from the previous book. Review it before seeking other documentation, and editorialize as you go along so you can write it better this time.

Ask your prospective interviewees for memos, specs, flow-charts, anything pertaining to the subject.

What do you expect the next four paragraphs to be about? In what order? Are they?

Memos belong in the progress folder.

Specifications contain some valuable information: what the product is to be from the user's viewpoint, which is often the developer's view of the user; how the product will interact with other products; and many of the terms, which are the names of product parts. Specs generally outlive their usefulness by the midpoint of the project though, because they become out-of-date, therefore inaccurate.

Some programmers actually do draw flowcharts before they begin coding. If you will be writing an internals book, then flowcharts are an invaluable aid. If the book is to be an opera-tors guide, then flowcharts are the wrong view for the user, but they may be helpful for your understanding of the overall project.

Sometimes you get lucky because there are transparencies from a presentation, or better yet, a videotape. The tape will not only introduce you to the subject, but give you a sneak preview of the person you'll be likely to interview.

Photocopy all the hardcopy and transparencies so that you can highlight parts that seem important and put question marks next to parts that seem inexplicable.

If there is nothing of value available yet, then look up articles on the subject in trade journals. There is nothing totally new under the electronic sun.

But woe to the writer who works from documentation alone. The day that a memo was sent from the engineering department saying that they were changing the product, the marketing department said they could not sell it, the packaging department said they could not get the parts for it, and the lawyer said the company was responsible for special government clauses. Did you get the follow-up memos?

17.2 INTERVIEWS

Interviewing is the most critical—and often the most exciting—part of technical writing. Your company is paying for some of the best minds from all over the world, and they are your teachers.

Take best advantage of your interview time and the information you get.

17.2.1 Prepare

To make the most of your interview time, prepare for it.

17.2.1.1 Schedule Time

After having established a rapport with the interviewee (often the product originator and/or developer), your dropping in to watch the program run or the hardware being tested or your calling to ask a quick question over the phone is efficient and generally welcome. At the beginning though, establish your requirement to set up official time for interviews.

Scheduling a conference room is best because interruptions are less likely. If there are crunches for conference rooms at a time convenient to both of you, then your company cafeteria at off hours or patio tables—on nice days—are alternatives.

A one-hour time slot is the norm. Most industry habitués automatically begin pushing their chairs back, closing their notebooks, and checking their watches at the end of 55 minutes, much like students.

17.2.1.2 Take Props

Using any or all of the following props makes the information that results from the interview that much more preservable.

1. *Questions.* Prepare a list of questions based on your reading. Keep them general rather than focused on details,

although at this point, it's not always possible to tell the difference. You may also want to give your interviewee a copy of the questions ahead of time.

2. *Paper and Pens.*   Have plenty of paper and different colored pens so that your interviewee can explain by drawing diagrams, which seems to come naturally to so many programmers and engineers. It is much easier to carry a sketch back to your desk than to try to hurriedly copy information off a blackboard—if one is even available in your meeting place.

3. *Notation Standards.*   If engineers or programmers supply documentation, give them a copy of the notation standards you use: spelling list, command notation table, and format examples. It'll save some time in incorporating their portions in your book, and it'll salvage some consistency when you have to incorporate their portions verbatim because of last minute changes to the product.

4. *Tape Recorder.*   You're writing about electronics; use them. Tape recorders are getting better and cheaper all the time. Have you ever conducted an interview and felt that, indeed, you understood the product—until you walked back to your desk to write about it?

    Of course that list of questions will help, but how many words can you write a minute? Those scribbled answers that you were sure would jog your memory, do they start to look like "Jabberwocky" now? How about sketches? What do those arrows mean again?

    A tape recorder offers more than just the advantage of accurate recordings:

    1) It allows you to look at the speaker rather than to be scribbling answers.

    2) It keeps you and your interviewee on track, less likely to wander from the subject.

    3) It records your spontaneous questions in cases where the interviewee's answers are unexpected, incomplete, or incomprehensible.

It is generally better to use battery powered recorders so that you need not be constrained as to where you meet by the length of a cord. Take extra batteries along. Also, be aware of noisy air conditioners inside and noisy power mowers outside. Both can cause frustration when it's time to transcribe the tape.

17.2.2   Conduct

One of the most delicate parts of a tech writer's job is conducting the interviews to elicit information that can be found only in the mind of the engineer or programmer who has created the product you must describe.

Everyone working in hightech is under pressure of a deadline, whether self-imposed, carrot-led, or boss-dictated. Your interviewee may have just got a speeding ticket on the way to work, and/or a new boss who is planning to reorganize, and/or the flu.

Your interviewee may be just out of school, with EE or comp sci diploma still curled. Others have been programmers or engineers for 15 or 20 years, learning their craft on the job, coding in 1s and 0s or drawing without light pens.

Some came from established companies—IBM, GE, the BUNCH; others, from hundreds of small specialized suppliers. Each is accustomed to the ways business was done there, ignoring or unaware that it's supposed to be done somewhat differently here.

You may be in one or more of the same boats.

Though you may have read *The Psychology of Computer Programming* or the *Existential Pleasures of Engineering*, so far no one has written *The Existential Psychology of Tech Writers*, so the burden to understand—and to be understood—is on you.

17.2.2.1   *Guide the Interview*

Some interviewees are well-organized writers and speakers; others may be very good at their work but little able to describe it. Gauge your participation and control in the interview dependent on the interviewee, and guide the interview accordingly.

When people you interview speak easily, let them go. Jot down unclear words and phrases when you hear them, and wait for a pause to get clarifications.

When people you interview are hesitant, asking a question from your list can get them talking. If that doesn't work, then start discussing their design yourself. They will correct you soon enough.

Never be afraid to ask a question about an explanation that you don't understand the first—or second—time. Don't worry about seeming uninformed; it may work to your advantage, making the interviewee more comfortable. And you may uncover an error or help clarify a development problem for the designer.

Ask open-ended, not yes-and-no, questions. What is the purpose of the code, the piece of hardware? How will it help the system? The user?

Convince your interviewees that you care about the success of their work, that it will work better because of your work. In order to make them believe you care, you have to care. If you don't, then you must be pretty dissatisfied with your job—or temporarily burnt out. Plan to do something about it—as soon as the book you are writing is finished. A light at the end of the tunnel will help.

17.2.2.2 Define Their Terms

Every interviewee defines acronyms—that's easy. It's terms like *module*, both in hardware and software, whose definitions are the functions, the components, the structure of the product that have to be precisely described.

Did you notice that the header was not repeated in the last paragraph? Is the term "precisely described" definitive enough?

People are often surprised to find out that other people sitting three desks away are using the same term to mean something different. Don't expect to change anyone's use of a word, but you will be able to define so that it's clear in the context of your book.

17.2.3 Transcribe

Write rough paragraphs about any drawings done in the course of the interview, and read over notes while they are still warm.

If you have taped the interview, then use a transcriber: you can regulate the speed of playback and control operation with a footpedal, so the transcriber is much more efficient than a tape recorder.

If you input the information at the terminal yourself, then you will probably not transcribe word for word but edit as you go along.

If you have data entry personnel who act as inputters, then help them understand what you need. Though you will not necessarily want comments on the weather by you or your interviewee transcribed for posterity, it's better than missing important information, and remember the inputter may not have had much experience with terms particular to the company products.

Be prepared; what sounded like a lot of good information during the interview may well take some ferreting out from tapes that are transcribed word for word, like the following:

> There isn't time enough in this meeting to go through these all in any kind of significant detail, so what I have thought to do is to give you a general overview of the operations and their relationships and if we don't get to mention all of them then we'll take it up at another meeting. Is this going?

If you said, "Yes, it's going," then asked the interviewee, "Please name all the operations," at this point on the tape you will be pleased to hear the operations listed. Transcribed and edited, they will provide the introduction and enumeration of sections in your book.

If you let the interviewer go on much longer without asking to have the operations named, you may get:

> In the time we have today then I'll start with the operations that move the data onto disk and depending on the time remaining we'll see how many we can get to and they're really not all actually completely defined as to details at this time. . . .

Depending on the convention you and the inputter observe for undecipherable words and phrases, usually question marks, the transcript may look like this:

> If there exists a record in the ??? such that the identifier and the ??? length match in the database, then the record is read into the ??? processor and. . . .

Even words that the inputter may be accustomed to seeing on printouts are not so easily understood on tape.

Whether you or someone else has done the transcription, listen to the tape again while you read along from the hardcopy; that way you can check the input and reinforce the information.

Once you have edited away the unusable and turned the conversational phrasing into more appropriate prose, the hardcopy serves as a first rough draft for sections of the book.

17.2.4 Return

Like most processes involved in electronics and in tech writing, interviewing is iterative. After each interview, you will return to your desk with more material, more clarification, more detail, and most assuredly more questions.

If one interviewee cannot supply all the answers, then do not hesitate to ask for other sources.

17.3 HANDS-ON

Participate in any hands-on situations you can.

If you're writing about a piece of equipment, then look at the real parts, not just the drawings.

If you're writing about an application or a system, then use as much of it as you can.

If you're writing about an installation or maintenance procedure, then watch it.

If you're writing theory of operations, then learn about the architecture and the rudiments of reading data flow diagrams.

All products need to be tested. If you've just begun the book, then observe the testing, or, better yet, be a test participant. It's a good way to see the product from your readers' view. If you've already written quite a bit of the book, then ask for it to be tested along with the product.

17.4 CLASSES

If you are lucky enough to be working for a company that holds field and/or customer training courses and are writing about an existing product, then enroll in the course. It's a great chance not only to learn about the product but also to meet users of the current publications.

If your company doesn't have classes, maybe you can attend a seminar on the subject—especially if there's one in your area. That's a good reason to open some of the "SORT" mail: we all get promos for seminars and books.

17.5 CONVERSATIONS

Talk with people who know something about the product.

Talk with other writers who have worked on similar projects or who are sharing the assignment. When you hear something they might need for their books, let them know, and ask them to do the same for you. Good writers are always gathering information, and sharing it helps all and leads to better publications.

If there's a training and education department, then talk with the instructors. They will probably want to use your book in planning courses and are themselves learning about the product. Again, share information.

If you are working on a revision because of product enhancements and you get the opportunity to visit a customer site, then talk with users of the product. That can be the best information of all.

18

WRITING

Now that you have started a folder for the project, collected stacks of documents, notes, and illustrations, and talked to people, it's time to begin writing in earnest. Unfortunately, it is not true that inspiration enters the writer's head, travels through the arms to the pen, pencil, or keyboard, and results in paper covered with concrete words, figures, and tables.

As Martin Luther said,

> He who does not know writing thinks it to be no work. Three fingers work (that was before word processors) but the entire body is at work. It is great work, commendable in him (also before feminism) who does it for the right reason.

Our reason is to help the people in the audience learn something they need to know.

In order to help, you will be providing formal organizations containing the information you have gathered and will continue to gather and refine throughout the process of constructing the book.

18.1 ORGANIZE

The people who will read the finished publication will be reading it to learn or paging through it to recall something about the system. Understanding requires order. Organization helps the readers by setting up the overall patterns of expectation.

How did you learn to understand the product? Was it difficult? do you think there would be an easier way to learn? Organize your book that way.

The scientific procedure for solving problems fits the organization process very nicely: First, try to arrange all the information; then try to strike out all the irrelevant information; and then try to see if there is some new way to put all the pieces together to form a different picture.

Don't hesitate to form a different picture; in other words, you may need to reorganize the book more than once before you reach the optimal tradeoff.

1. *First categorize the material.* Look over the transcripts, and group the parts that seem to best fit together. Use any previous publication to help group them—unless you've come up with a better way.

2. *Next, outline the publication.* The principle of outlining is arithmetic: division and addition. No level can be divided into fewer than two parts, but you may add as many parts as you need, and the parts do not need to add up to an even number.

   Name the topics into which your groups of information fit, and then arrange the topics according to their position in the hierarchy or sequence you will be describing. This search for structure will almost always yield discoveries about the product and show what's still missing.

   When you have finished the outline, you have constructed your first table of contents, which is a summary for the current logical structure of the book.

3. *Keep the table of contents current.* If your text processing system generates the table of contents automatically, as it should, then the table will change whenever you change the organization. If you have to update the table of contents manually, then it is worth the effort because you will be able to find old information and place new information more easily.

4. *Keep track of the information.*   If you are writing about a large system that needs more than one publication and some of the information is needed by more than one audience, then you can either repeat the information in each publication in the complement—tailored to the audience for that publication—or refer the readers to the other publication. In general if the reader needs the information to perform the task you are describing, then try to include it. If the information is not essential to the task at hand, then refer to it.

18.2   WRITE

Now that the programmers are producing lines of code and the engineers are ordering parts to build the prototype, you will begin to write text and draw illustrations.

1. First, write down all the information you have gathered into the organization you have outlined. Don't pay much attention to rhetoric at this point—get it down.

2. As you read or hear new terms, add them to your glossary list.

3. As you continue to gather more information, begin to evaluate it for elaboration or exclusion.

4. As you continue to write, draw components of the product to help you to see them more clearly. The drawings will probably form the basis for figures in the publication.

   Even when you have an automated graphics system, even when the technical illustrator is part of the department, even when you create illustrations electronically, it takes time, so you'll want to get the figures started as soon as possible. When you've sketched an illustration, give the drawing to the developer to critique. Make the appropriate changes and turn it into artwork according to your company's protocols.

5. As you write, continue to watch the prototype being built or the software in action.

6. As you understand the product more, begin to exclude information your audience does not need to know. Remember, the information you get from the developer is seldom from the point of the view of the audience. The developer may be very proud of a piece of code that provides a particular feature, but it's very unlikely that the audience will care about why it works—they'll care about how to work it.

7. As you write, reread what you've already written to catch information that is incomplete.

8. As you write, continue to review the table of contents to be sure that the organization is optimal. Ask another writer to tell you what the product is about and what audience it's for by looking at the table of contents. If the answer is right on or close, then you've established the purpose and audience clearly.

9. Write the preface. Does the text really match the purpose and audience? Is the organization logical from the reader's point of view? Do any pre- or corequisite publications need to be mentioned?

10. As you approach the time for a technical review, check the draft for consistency of terms, logical constructions, and typos, but you are too close to the prose now to do a thorough edit, and the reviewers will make changes, which you will then have to incorporate.

NOTE: If you have scheduled time for only one formal technical review, then give the developer copies of the draft in chunks of a section or more as you are writing. These previews will help you both to avoid surprises at the full review.

19

REVIEWS

Before your book is published, the drafts may be reviewed by developers, managers, marketeers, engineering change boards, quality controllers, and lawyers, all of whom won't agree with some parts of the book—or with each other.

The review drafts will have to be prepared, distributed, collected, and incorporated in one draft before final editing and production can begin.

Were you surprised to see the asteriks here? Why not?

19.1 PREPARATION

To prepare the draft for review, print off the current copy, including the preface and appendixes.

Take a long look at the table of contents to see if the organization still seems the most logical. If not, then move the sections around now.

Proofread the text for gross grammatical errors and format inconsistencies that obscure meaning. Include drawings that you've planned for figures.

If this review is to be the first of two, then you probably will have quite a few questions about the material. Place the questions in text and highlight them by bold, caps, or underlining followed by question marks, for example:

ARE THERE 4 OR 6 CCAs???

Remember the product is probably still changing, so ask some open-ended questions too, for example:

DID YOU ADD ANY MORE SCREENS???

If this review is to be the final review and you still have questions, then make sure to call attention to them: bold, caps, underlining, and arrows all at once usually does the trick, for example:

$- \rightarrow$ **IS THE CHOICE ON A MENU OR A FORM???** $\leftarrow -$

19.2 DISTRIBUTION

Once the draft is prepared for review, print off as many copies as you will need for reviewers, and attach forms telling what you expect of the reviewers and by when.

A general review approval form goes to all reviewers who have a say in publishing the book. In addition, attach a note to technical reviewers' copies with specific requests as to how to mark the draft.

NOTE: Both review forms should be boilerplate so that you don't have to compose them at each review. You can always add extra information specific to a particular book.

19.2.1 General Review

The general review form goes to all reviewers designated in the publications plan. Just fill in the blanks. Figure 19-1 illustrates an example General Review Form.

To: *(usually "Distribution List")*   From: Technical Publications
                                            Date: *(of distribution)*
                                            Subject: Publication-in-Process

The attached publication draft,
_____ *(publication name)* _____ , is in

_____ preliminary review

_____ final review   *(check one)*

_____ update/revision

Your copy is due back on _____/_____/_____
           (no more than 2 weeks from the current date)

Please review the copy, check your preference, sign below, and return.

_____ I approve; continue to next phase.

_____ I approve upon incorporation of the specified changes.

_____ I withhold approval until resolution of major changes.

                              Signature: _____
                                  Date: _____

Please list additional reviewers: _____
                                            _____
                                            _____

Figure 19-1. General Review Form

19.2.2 Technical Review

The technical review note goes to all technical reviewers, usually developers, designated in the publications plan. Figure 19-2 illustrates an example boilerplate for the note to technical reviewers.

19.3 COLLECTION

Review drafts and signoffs should be back no more than two weeks after you have distributed them. If they are not returned in time, then they are probably at the bottom of a stack of listings or under a pile of engineering drawings. It is your responsibility to collect the reviews.

If a review copy is a day or two past due, then call to remind the reviewer and ask when it will be returned. If it becomes more than three days late, then take your problem—and it is definitely your problem—to your boss, the reviewer's boss, or whomever you think will have the solution—and the clout to implement it.

When you receive each technical review copy, first check to see that all your questions in the draft have been answered.

If you receive copy that comes back unscathed, then beware. The technical reviewer, already expert in the material, therefore not really reading for information, may easily have missed ambiguous explanations or incomplete instructions. Technical reviewers—usually programmers or engineers—are busy too, as they oft point out, and a thorough technical review takes several hours minimum. The reviewer may have merely found a typo or two, signed off the general review form, and moved on to make more design changes in the product or ferret out a particularly elusive bug. Ask to do an hour's review together. You will probably find an error or two; if not, then congratulations—you've written the first perfect book in my experience.

For the books that need legal authorization, generally hardware manuals that describe potentially dangerous conditions, be sure that you have got the authorization before going any further.

TECHNICAL REVIEWERS PLEASE NOTE:

This review draft is being submitted to you because you know whether the information is accurate and complete.

When you encounter question marks, an incorrect statement, or a confusing explanation, please answer, correct, or clarify.

When you make changes, cross out the words that you want changed and write in your correction. Please write legibly and *use a red pen* to make technical corrections.

If you wish to comment on organization, style, or general content, feel free to do so—in a different color pen.

Ignore format inconsistencies, as they will be corrected after incorporation of changes, but if you catch a misspelling or a typo, then mark it, of course.

If more than one person is to review a portion of the draft, then please review sequentially, stacking comments on one copy.

When you've completed your review, sign and date the attached approval form and return it to:

(name of appropriate person, usually writer).

If you have any questions, then call _____ or _____.
(extension numbers of publications personnel).

Technical Publications Department
Mail Stop C123

Figure 19-2.   Note to Technical Reviewers

19.4 INCORPORATION

After you have collected all the review copies, sift through them for obvious problems.

If you receive general review forms withholding approval, then call immediately to begin resolution of the changes.

If you receive conflicting comments—as is almost always the case at least a couple of times in each book—then it might seem that the best way to resolve the inconsistencies is by a meeting between you and the disagreeing reviewers. However, a meeting among more than two reviewers will rarely end in consensus, so you may be better off speaking to each independently and suggesting that they get together, but be prepared to act as messenger in order to resolve the difference yourself.

After the approval and technical comments have been resolved, incorporate the technical changes in the draft.

Don't feel obligated to implement all other comments without question, but respond to them all. If it's a good suggestion, then make the change—and remember it for future publications. A diplomatic way to handle rhetorical suggestions that you find unacceptable is to tell the reviewer that you cannot make the change because of consistency considerations.

If the review was preliminary, then begin the writing and review process once more. If the review was final, then it's time to begin the technical and mechanical editing.

20

EDITS

If there is a universal view about good writing, one that all rhetorical and technical writing texts agree on, one that most writers at least pay lip service to, then it is that EDITING IS ESSENTIAL.

Good editing requires continual practice. Luckily, while you're practicing, you're getting it done.

The more you write, the more you recognize the need for editing. A few tech writing groups include a technical editor; some include copy editors, but very few writers have time for peer editing. If you don't do peer editing as SOP, then at least make time to swap the introduction to the book with other writers. In that way, you can help each other to start out on the right foot, teach each other writing tricks, ensure greater consistency among the publications, and learn something about another product or another view of the product you're writing about.

If you are editing your own writing, then don't begin for at least two days after you have incorporated all the technical changes. You've usually got more than one project going at a time, so that should be easy.

No matter who is responsible for what kind of editing, there's only so much time, so keep each operation in perspective. Style is important, but the fine points that writers love to argue are less likely to detract from your readers' understanding than poor organization or inconsistent diction. Of course the accuracy of the content comes first.

In each edit:

1. Check content.
2. Check organization.
3. Check style.
4. Check references.
5. Check notation.
6. Check consistency.

Numbers 1, 2, and 3 above are discussed under technical editing; numbers 4, 5, and 6, under mechanical editing, as if they could be separated and labeled so neatly.

So that all editing is consistent, choose a subset of the standard sets of copy editor and proofreader's marks found in any standard dictionary for your department to use. Copy/proofreaders marks are yet another level of notation, a kind of metanotation used to make editorial comments about the book. Table 20-1 is an example subset of copy/proofreader's marks with meaning, example, and result of each.

I could have included the table on copy/proofreaders marks under "Notation." Would that have been a better choice because of consistency, or is this a better choice because of relevancy?

Some pubs departments are becoming automated to the point of using terminals to do editing. Phrase highlighting, notes in windows, and color coding replace proofreader's marks in these systems.

20.1 TECHNICAL

A technical edit includes the technicalities of both content and rhetoric. You can generally find at least one unclear, ambiguous, or incomplete description—no matter how many technical

Table 20-1. Copy/Proofreader Marks

Symbol	Meaning	Example	Result
⋏	insert	changing⁄phrase *(this)*	changing this phrase
↶	delete	changing this ~~phrase~~	changing this
⌣	close up	chang ing this	changing this
⌐	move left	⌐changing this	changing this
⌐	move right	changing this ⌐	changing this
⌣⌐	transpose	this changing	changing this
¶	paragraph	¶ Changing this	Changing this
△	blank character	changing△this	changing this
≡	capitalize	changing this	Changing THIS
=	small cap	changing this	CHANGING THIS
⌐ /	make lowercase	CHANGING THIS	Changing this
(stet)	leave as it was	[not changing]	not changing
#	line space (below)	# changing this	changing this

reviewers have checked the document. That's why you have to understand how the product works as well as how the language works.

Ideally, the publication would get three full technical edits: The first for content and completeness, the second for organization, and the third for style, which is everything you missed the first two times.

Three passes are the ideal, but so are absolute zero and the perfect publication. More often there is barely time for one, so you will be looking for content, organization, and style concurrently—and in that priority.

20.1.1 Content

Read the book for sense. Note any rhetorical errors quickly, but pay special attention to the logic in description. Is it complete?

Checking what may be left out is as important as checking what's included.

For a simple bad example:

[Exchanging a channel card includes:

1. Unlocking and opening the access panel.
2. Unscrew the bolts from the corners of the front cover and remove it.
3. Slide off the safety clip in front of the card to be replaced.
4. Slide out the card to be removed.
5. Slide in the replacement card.
6. Replace the front cover and tighten the bolts.
7. Close and lock the access panel.]

The rhetorical errors are obvious, right? First, there's a lack of parallelism between step 1 and step 2, which switch from declarative to imperative. Second, "the card to be replaced" and the "card to be removed" refer to the same card, so it should have the same name. But did you notice, the writer forgot to replace the safety clip?

In Step 2, would you use a plural pronoun if there were more than one cover? Try it. Now, does "them" refer to the covers or the bolts? It would be better to repeat the plural noun than to use a pronoun, for example: **Unscrew the bolts from the corners on the front covers and remove the front covers.**

20.1.2 Organization

By now the overall organization of the book should be solid, but if the technical review included major changes, for example, a feature added or deleted from the product—it happens even at the last minute—then the organization might be affected to the degree that you deem reorganization is in order. Sometimes you just have a better idea. In either case, call your reviewers and propose the reorganization before implementing it.

It is more generally the case that internal structures need to be reorganized. Read over the rules in Section 11.2 and the structures in Section 11.3 as reminders. The organization should be the best you can make it in the limited time.

1. Check for correct parallelism at all levels throughout the book.
2. Check sentence structure:
    1) Check that each conjunction is appropriate.
    2) Check that each pronoun has and matches its antecedent.
    3) Check that all modifiers are correctly placed and punctuated.
    4) Check that each use of *this* is followed by a noun.
    5) Check for strings of adjective/nouns that can be clarified by a hyphen or a prepositional phrase.
    6) Take a look at sentences over 25 words, but consider each clause in a compound sentence as one sentence.
3. Check paragraph structure and organization:
    1) Check that paragraphs describing an operation are in the correct sequence.
    2) Check that paragraphs describing parallel parts of a machine or program are parallel in structure and phrasing.
    3) Split paragraphs of more than ten lines.
4. Check that introductions, transitions, and connectives make relationships clear.
5. Check the validity of references. Be sure that any figures calling out objects and locations really match and illustrate the descriptions in the text.
6. Finally, check the table of contents once more, and make sure the information in the "ORGANIZATION" section of the preface is consistent with the table.

20.1.3 Stylistics

Read over the list of phrases in Table 12-1, "Economy," and the pre- and proscriptions in Chapter 15, "Style," for reminders. The style should be the best you can make it in the limited time.

Be flexible: There are generally at least two good ways of writing something. Don't automatically strike out redundancy, passive voice, expletives, or a long word.

Sometimes redundancy is acceptable—especially when it stresses an important point that the reader might miss.

Sometimes the grammatical subject is not important; it's the object of the action that needs discussion, so passive is the correct voice.

Sometimes avoiding an expletive casts a sentence into a more awkward construction.

Sometimes you really mean "initiate" and not "start" because "initiate" is the name of a specific function or command.

20.2 MECHANICAL

A mechanical edit, sometimes called a copy edit, includes checking for accuracy in matters of form—references, notations, and consistency—throughout the publication and among publications in a complement.

Copy editing requires, above all, close attention to detail. If you are copy editing your own book, then remember that it is especially difficult to catch your own mistakes. If your department includes a copy editor, then encourage their mentioning any errors missed in the technical edit.

20.2.1 References

Inaccurate references will damage the integrity of the publication for any reader who happens to spot them.

1. Check internal references against the figure, table, section, or appendix that they refer to, and make sure that they match.

How would you recast the above sentence to get out of the *they* reference? Is the sentence clear even though the first "they" is separated from its antecedent by a series and the second "they" has no grammatic antecedent but is the result of a match between a figure, table, section, or appendix and its internal reference? Does that help clear things up?

2. Check that the title and publication number of each external reference matches the actual book.
3. If you don't have a system that automatically numbers headers or lists, then check the sequence of each by just looking at the string of numbers—don't get caught in the text.
4. Check topics introduced by a list to make sure the items are discussed in the same order as listed.

20.2.2 Notation

Use Chapter 3, "Notation," and the department standards to correct errors in command presentation and text punctuation.

1. Check that all explanations of fields in a command match the command syntax.
2. Check punctuation, making sure
    1) That a comma does not take the place of a colon, semicolon, or period;
    2) That a series of three or more parallel structures takes a comma or a semicolon before the *and* or *or*; and
    3) That apostrophes are always and only where they're supposed to be.
3. Check that each acronym (in parens) immediately follows the first use of the word. Be sure all acronyms are listed in the glossary.

20.2.3 Consistency

Consistency enforces patterns, which enforce learning. Enforcing consistency may be tedious, but checking for it is even more so. However, like with any editing, repetition improves the quality of the book and eases the task.

1. Check spelling. Use your department spelling list to check for consistent capitalization, hyphenation, and abbreviation.

    If some words in the book have possible alternative spellings, for example, *sizable* and *sizeable*, then be sure the words are not spelled differently on different pages. Choose one spelling—and add it to the spelling list. (The first entry of acceptable spellings in the eighth edition of *Webster's New Collegiate Dictionary* is a common standard.)

2. Check that each set of commands and error messages is presented in consistent format.

3. Check the headers and figure and table titles for correct capitalization.

4. Check alpha and numeric representations and symbols. Do not alternate, for instance, between *ten* and *10* or *percent* and *%*.

21

PRODUCTION

Producing the publication includes preparing and proofreading the final text, following all format standards for each component, arranging pages logically and aesthetically, putting the master together in the correct sequence, and sending it off for reproduction or typesetting.

21.1 TEXT

Preparing the final copy of text is the last chance to catch errors.

1. First add all the changes from the technical and mechanical edits to what will become the "new copy."

    If you have done neither of the edits and any of the editing marks are unclear, then ask. Don't guess.

    If you find an error in spelling, an acronym that has not been defined at first use, an inconsistency in notation or spelling, then mark the correction on the edited copy and include it in the new copy.

2. After the new copy is complete, proofread carefully. Read from the original manuscript to the new copy, not the other way around.

Try to read each section independently, starting from the last sentence in the section back to the first: you are less likely to get caught up in the text and more likely to catch errors.

If you have the time and the personnel, then have one person read word-for-word, punctuation-mark-for-punctuation mark, from the new copy while the other person notes any errors or omissions.

If you read the new copy from the terminal screen, then you can make any corrections immediately.

3. If you do not have automatic table of contents generation, then spot-check page, section, figure, and table numbers and titles once more—even when you haven't changed any of them—to be sure that the book matches its table of contents.

21.2 FORMAT

Once the text of latest new copy is correct, determine the size of the figures, leave space for each figure, and then arrange each page of text both logically and aesthetically.

1. Use white space to set off logical divisions of information.
2. Don't leave widowed or orphaned lines, especially in a list.
3. Don't leave just two or three lines on a page unless it is the last page of a section.
4. If the publication requires special running heads or feet, include them.
5. If the publication is to be contained in a binder, allow enough margin space to punch holes.
6. Paste the final artwork into the appropriate spaces. Be sure that in all the tables and figures in which two lines join at an angle, the lines meet perfectly.

You now have a final copy.

21.3 COMPONENTS

Use Chapter 1, "Front and Back Matter," or your company's equivalent boilerplates to prepare the standard components.

1. On the title page, fill in the following:
    Title of manual
    Publication number
    Date
2. On the back-of-title-page, fill in the following:
    Copyright notice date
    Revision notice
    Trademark notice
    Reference to reader comment form
    Disclaimer
3. On the revision history, fill in the following:
    Title
    Publication number
    Date

If you include a table of page numbers on the revision history page at first release, then create it now.

4. On the reader comment form, fill in the following:
    Title
    Publication number
    Date
5. If you have to create the table of contents manually, then fill in the following:
    Headers through the third level
    Figures
    Tables
    Appendixes
    Glossary

21.4 MASTER

Each page of the book under production should be printed one sided for the master. Check over all the pages—text, format, and components: The cleaner you make the master, the more professional the book will look.

Arrange the pages of the publication master in the following sequence:

1. Title page
2. Back-of-title-page
3. Preface
4. Table of contents
5. Frontispiece (optional)
6. Sections
7. Appendixes (optional)
8. Glossary
9. Index (if possible)
10. Revision history
11. Reader comment form

After arranging the components, fill out any forms required in your company, for instance, a request for covers or binder inserts (they have already been printed because you notified the production staff way back in the planning stages), a reproduction request, or a pagination sheet.

Then check the master one more time, slowly, and deliver it for repro or typesetting.

Now how are you going to celebrate?

PART IV: MAINTENANCE

The Technical Publications involvement with a project doesn't end with the delivery of the publications to the printer. Just as the development group must maintain the product, the publications group must maintain the manuals that go along with the product.

Publication maintenance includes archiving released publications and, hopefully, revising and/or reprinting them.

The previous sentence contains two structures on which grammarians, professional and amateur, disagree. Do you recognize them? I accept "hopefully" even though it has no grammatic referent—and in spite of the publisher's objection. I applaud "and/or" because it is clearly logical and shorter than "both or either."

By the way, how do you represent "both or either" in computer logic? Does a 2-input OR gate fit the requirement? Logical symbols are not necessarily the same in different systems.

The reason for "hopefully" is to signal anticipation of positive results. Revisions mean that the product is a success: New features are being added to the hardware, the software, or both.

Revision also means the chance to improve the manual. The IBM *Principles of Operation* is at level 9 and has improved with each revision. A publication, like a program, needs a shakedown stage where bugs that weren't caught during development or check-out may be found by users.

Reprints mean that the product is a success: there are more customers buying the product than there are copies of the publications.

22
ARCHIVES

By the time a book is finally released to its readers, the writer is usually enmeshed in a new assignment and barely glances at the publication when it's delivered—unless, of course, it's the first or maybe the second one you've written.

NOTE: If you store your publications on disks in the company's data center, then do you know if and after how long files that are not accessed are deleted? Be sure that you read memos from the data center.

Depending on your company's procedure, the master for the publication may be kept by document control or by the pubs department, and the text is probably stored on disk and eventually tape. If reader comment forms come in, they are looked at, answered, filed. And that's that. So the usual maintenance procedures don't teach you much.

Instead of the usual procedures, keep an active archive file for each publication.

The archive file for each publication should be available to all members of the department, and it should include: hardcopy of the publication, the distribution list and all signoff sheets from reviews, any returned reader comment forms and copies of answers to the comments, and notes and memos about the product's status—all in one place.

If the product is a one-shot, the archive file will serve as history. If the product is to be enhanced, then you have a head start on the revision.

22.1 COPIES

Keep at least two printed copies of each publication. Mark one copy "For Reference"; mark the other copy "For Learning."

Take time to look at the learning copy, whether you or someone else wrote it. Continue to edit it, technically and mechanically. You will see unclear descriptions, style deficiencies, and typos. If it's your book, then they'll hurt more, but press on. Look at the print quality and layout of the pages. Is there enough white space—in the right spots? Are the illustrations uncluttered? Do their titles match their subjects? Mark the corrections as you would for any edit, using the department standard spelling list and copy/proofreader's marks, and replace the copy in the file for others to edit.

When all of the writers have taken a look at the copy, use it at staff meetings to plan new publications, to change any format and notation standards that you see don't work as well as you thought they would, and to learn from your own and each other's mistakes.

In Silicon Valley, where everyone seems to move one company to the right every 1½ to 2½ years, the originator of the book may well have left, but get in the habit of discussing the copy in terms of positive and helpful corrections even when the writer is not present. Everyone is trying to improve all publications, not pick out one writer's errors.

If you are the writer who produced the book, then first you have to make a concentrated attempt to distance yourself. It may help, up front, to discuss the problems you had writing it so that you won't feel compelled to defend every correction. Then talk about the publication's history. Was it a new subject to you or one about which you had written before? Were your interviewees helpful or recalcitrant? Were reviews back in time?

How many changes were made to the product after you became involved? How difficult was it to organize the material? Why? If you could change one thing in the publication, what would it be?

If you are not the writer who produced the publication, then don't nitpick. First, what was good about it? Then, how familiar were you with the information before reading the copy? Was it easy for you to understand the overall structure of the product from the point of view of the readers who were cited in the preface? By the way, how do you perceive their point of view? Did you understand what they were to use the publication for? Might there have been a different way to organize it? Were the descriptions understandable? Logical? Not enough information? Too much? Did you use the glossary?

One person, maybe the originator of the publication, takes responsibility for taking notes about the suggestions.

If someone comes up with a better technique for a description, a better way to illustrate a point, a better way to write or format something, then make a note of it. If there is majority consent on a format or notation change, then incorporate the change in all the books you write from now on.

When you have finished, return the copy and the suggestions to the file.

22.2 DISTRIBUTION LISTS AND SIGNOFFS

Keep the distribution list and signoffs; they give you the names of the people who contracted for the publication and/or developed and tested the product. Contact a few of them. Ask for their reaction to the publication now that it's in print. If there are any in-house users, then ask them for comments.

The distribution list and signoffs will come in handy in case the publication needs revising, and you will have representative titles for gathering information even if the original people have moved on. (The signoffs may also be handy in cases of contention about technical errors that slid through.)

22.3 READER COMMENTS AND REPLIES

Keep and reply to all reader comments. Sometimes companies actually send technical writers into the field to talk to a publications' users. More often a few reader comment forms come back to the publications department. Consider all comments seriously; they are prompted by actual use of the publication to accomplish tasks.

Reply to all reader comments within a week of the time they arrive. Whether writing to a company employee, like a field service engineer in another city, or to an end-user who is using the publication to accomplish a job, you are doing PR for your department and your company as well as providing information.

Reply specifically to the comment, which is often a question or request for more information. If the reader asks something you cannot answer, then you may have to talk to the product developer, tester, or field service rep handling that account, but get the answer.

Have a simple reply form prepared, and fill it out. Since technical publications are not attributed to their authors and you may be replying about a book you didn't write anyway, it is not necessary to sign the form.

Figure 22-1 illustrates a boilerplate for reply to a reader comment. Of course you may have to add extra sheets for a longer reply.

22.4 MEMOS AND NOTES

Keep all information you may hear about the status and development plans for the product.

Ask to be put on the distribution list for memos that deal with the product and for announcements of training classes or demos.

If you have occasion to work with the same developers on another project or pass them in the halls, then ask about the product. Who's maintaining it? Do they plan to add any more bells and whistles? Make a note, and add it to the archive file.

Company Name (printed or in company logo)
Company Street Address
Company City, State, Zip

```
      TO:  (user who commented)
    FROM:  Technical Publications Department
    DATE:  (of reply)
 SUBJECT:  Response to your comment
```

Thank you for your comments regarding

_____ *(publication name)*

_____
(text of reply — add extra sheets when necessary)

_____

_____

_____

_____

_____

_____

_____

_____

_____

_____

_____

Figure 22-1. Reader Comment Reply Form

23
REVISIONS

A revision is the partial replacement of an existing publication because technical information must be added and/or changed. To revise a publication, you will go through the planning, research, writing, and editing operations, but in a shortened form.

You are now writing for at least two audiences: people who have been using the original product and publication and people who will be using both for the first time.

Make it as easy as you can for the old audience to find the new information and for both audiences to understand it.

You have a good start this time, an existing publication. Look over the publication, reader comments and replies, notes, and memos in the archives. If you will be revising a manual you wrote, then you can reacquaint yourself with the subject; if you will be revising a manual someone else wrote, then you will be doing research.

If the publication has not been edited since its release, then do a cursory edit. Don't expect to be able to get all the editing into the revision, but you can get some of the rhetorical improvements in with the technical changes, and it will be a better publication.

Be sure to follow all the standard conventions so that the new material will be consistent with the old, both technically and mechanically.

23.1 REVISION PACKAGE

A revision package includes the changed pages of sections and appendixes; an updated title page, preface, table of contents, glossary, index, and revision history; and a cover page.

In what order do you expect the following third-level sections to be? Have I made this point enough times that it will be easy for you to remember? Do you see why it is so important? Introducing the information (whether by series or by list) in the order that it will appear is not only an enforcing pattern for your readers, but also a constant check on the logical organization of the book.

23.1.1 Changed Pages

The changed pages contain any additions or deletions made to the text, figures, tables, and appendixes in the existing publication.

You may need to add change bars and extra page numbers for overflow in section or appendix pages. A text processing program comes in handy to mark the changes with change bars in the margin, but seldom can it mark the overflow pages.

23.1.1.1 Change Bars

Change bars are helpful to readers already familiar with the manual. Because change bars are meant to call attention to the latest changes, remove any bars from previous revisions before adding new ones.

Put a change bar at the beginning of each line that contains changed technical material, but don't put a change bar before lines in which you make rhetorical or typographical corrections.

23.1.1.2 Overflow Pages

Adding material often causes text pages to overflow. Add a period and sequential lowercase letters to page numbers to show that they were not part of the previous release, even if the pages fall at the end of a section. If new material has been added between pages *2-3* and *2-4*, for instance, then number the pages **2-3.a** and **2-3.b**.

If you call attention to the location of change bars in the preface, then maybe you can avoid this time-consuming renumbering scheme and merely number pages consecutively throughout the manual.

23.1.2 Title Page

Update both sides of the title page to reflect the current level.

23.1.2.1 Front

Make the following changes to the title page:

1. Change the publication number according to company protocol for revision levels, which will probably dictate adding or upping a revision level on the publication number. Changes made to a hardware manual usually require an engineering change order (ECO), and that number may also need to be added to the publication number.

2. Change the date to reflect the month and year in which the revision is to be distributed.

23.1.2.2 Back

Make the following changes to the back-of-title-page:

1. Update the revision notice to tell the reason for the revision, for example: **This publication has been revised to reflect**

new features added to the word processing and graphics capabilities of the Super-10.

2. Add the current year to the copyright notice, for example: **Copyright 1985 and 1986 by Micromecca Corporation**.

23.1.3   Preface

Make the following changes to the preface:

1. If you have included revision bars, then indicate the sections and appendixes in which they appear.
2. If you have revised the publication to the degree that its organization has changed, maybe added a new section, then update the "ORGANIZATION" portion of the preface, for example:

Section 2 includes two new commands for the new features, "Supertype" and "Wraparound."
Section 3 includes five new commands for generating multi-colored pie charts.

Product changes seldom affect the purpose of the publication or its audience.

23.1.4   Table of Contents

If the table of contents is not automatically generated, then update it to include any overflow pages and new appendixes.

If the table of contents is automatically generated, then add a note at the end about overflow pages; or, if you are using revision bars, you may be able to number the new pages sequentially within sections as long as you have mentioned the revision bars in the preface.

23.1.5   Glossary

Update the glossary to reflect the new terms and acronyms associated with the new information.

23.1.6 Index

If you included an index in the first release, then update it to reflect new subjects. If you didn't include an index and you now have the time and/or the automation, then create it.

23.1.7 Revision History

If you included a revision history page that listed all the pages in the publication in the first release, then updating it will be easy. If you included a revision history page that didn't list the pages, then list at least the new or changed ones now, according to your company's publication numbering protocols.

Add the new publication number and date to the original number and date.

Table 23-1 illustrates a sample updated revision history page that records two releases of publication.

If you didn't include a revision history page and the reader keeps the cover page, as you will suggest in the cover page itself, then the publication will still have a revision history.

23.1.8 Cover Page

Though the cover page is the first page of a revision package, it is the last page you'll complete, after determining how the content changes have affected the physical form of the publication.

Figure 23-1 illustrates a sample cover page. The fields are explained after the figure.

Supply the following information in the appropriate field on the cover page:

> UPDATE FOR: The title, publication number, and month and year of the previous release of the publication, the same as on the revision history page.
>
> CURRENT LEVEL: The title, publication number, and month and year of this release of the publication, the same as on the revision history page.

SUMMARY: The reason for the update, the same as in the revision notice on the back-of-title-page.

INSTRUCTIONS: Boilerplate.

OBSOLETE AND CURRENT PAGES: The now obsolete pages in the left column and the now current pages in the right column. The current pages should consist of the changed pages included in the revision package.

Table 23-1.   Updated Revision History

Revision History
Super-10 User's Guide
1234567-000, March 1985
1234567-001, June 1986

A list of pages in this publication and each page's current revision level follows:

Page	Revision	Page	Revision
title	001	3-1	000
ii	001	3-2	000
iii	001	3-3	001
iv	001	3-4	001
1-1	000	3-5	001
1-2	000	3-6	000
1-3	000	3-7	000
1-4	000	3-8	000
2-1	000	3-9	000
2-2	000	3-10	000
2-3	001	3-11	000
2-3.a	001	3-12	000
2-3.b	001	4-1	000
2-4	000	4-2	000
2-5	000	4-3	000
2-6	000	Appendix A	001
2-7	000		
2-8	000		

23.2   REVISION MASTER

A revision master is very like the master for a first release, just shorter. Each page of the revision master should be printed one

Micromecca Corporation Revision Package

UPDATE FOR: *Super-10 User's Guide*
                           1234567-000
                           March 1985

CURRENT LEVEL: *Super-10 User's Guide*
                           1234567-001
                           June 1986

SUMMARY: This publication has been revised to reflect features added to the word processing and graphics capabilities of the Super-10.

INSTRUCTIONS: To update your copy, remove obsolete pages and add current pages from this package.

PLACE THIS COVER PAGE AT THE BACK OF THE UPDATE PUBLICATION.

Obsolete Pages	Current Pages
title page	title page
i	i
ii	ii
iii	iii
iv	iv
2-3	2-3
	2-3.a
	2-3.b
3-3	3-3
3-4	3-4
3-5	3-5
	Appendix A

Figure 23-1. Revision Package Cover Page

sided, and the cleaner you make the master, the more professional the book will look.

Arrange the pages of the revision package master in the following sequence:

1. Cover page
2. Title page
3. Back-of-title-page
4. Preface
5. Table of contents
6. Changed pages
7. Glossary
8. Index (if possible)
9. Revision history

Check the revision master one more time, slowly, and, as with the master for a first release, deliver it along with any required forms for repro or typesetting.

A little celebration is in order.

24
REPRINTS

A reprint is the total replacement of an existing publication. A publication is generally reprinted when 50% or more of it is being revised because of changes to the technical information or because the manual needs to be replenished in the stockroom.

24.1 REVISION

A reprint because of revision replaces all previous releases of the publication.

After the review of the archives, the planning, research, writing, and editing operations, you will include all previous revision package material and prepare a full master.

The following is a checklist for the reprint of a publication that has been revised:

1. Include all revision package changed pages numbered sequentially within the publication.
2. Remove all revision bars.
3. Update the front and back of the title page.

1) On the title page:

(1)   Change the publication number according to company protocol for publication numbering.
(2)   Change the data to reflect the month and year in which the reprint is to be distributed.

2) On the back-of-title-page:

(1)   Update the revision notice to say, for example: **This release replaces all other releases of this publication**.
(2)   Add the current year to the copyright notice, for example: **Copyright 1984, 1985, and 1986 by Micromecca Corporation**.

4. Update the preface to incorporate the changes made by all revisions.

5. If the table of contents is not automatically generated, then update it.

6. If you include a revision history table in your publications, then:

1) Remove previous publication numbers and replace them according to company protocol for publication numbering.

2) Remove all previous dates and replace them with the month and year in which the reprint is to be distributed.

3) Under the date, add a line that says: **This release replaces all previous releases of this publication**.

4) Update the page table as you would for a revision. All the page numbers should reflect the current level.

7. Create the master as you would for an initial release:

1) Title page

2) Back-of-title-page

3) Preface

4) Table of contents

5) Frontispiece (optional)

6) Sections

7) Appendixes (optional)

8) Glossary

9) Index (if possible)

10) Revision history

11) Reader comment form

As you would for a first release or a revision, check the revision reprint master one more time, slowly, and deliver it along with any required forms for reproduction or typesetting.

Time for another celebration.

24.2 STOCK

A reprint because of low inventory means you've got a big seller.

If revisions have been made, then the revision packages should be incorporated into the publication before it is reprinted.

Though the tech writing department may be responsible for data processing necessary to integrate revisions, there is no writing involved.

Celebrate—you've got enough projects going on already.

BIBLIOGRAPHY

Nothing is more beneficial—or more enjoyable—for a writer than reading. Read in as wide a variety of disciplines and genres as possible: technology, philosophy, sociology, essays, fiction (science fiction is fine). Read the trade journals like *Datamation*, *Electronic News*, and *Computerworld*; business journals like *Fortune* and *Harvard Business Review*; and popular science magazines like *Scientific American* and *National Geographic*.

Watch the language change. When was it that "arcane" and its follow-on "arcana" replaced "esoteric"? When did "diametrically opposed" become "two sides of the same coin"? What is the currently popular sentence structure, loose or periodic?

The books in the following annotated bibliography lists progress from the basics of technical writing to the heights of writing technical. You can refer to them for grammar and format, writing and style, reference and technical research, and for real pleasure.

I. GRAMMAR AND FORMAT

The following books deal with the basic mechanical skills that every good writing department needs.

American Heritage Dictionary of the English Language: New College Edition. Boston: Houghton Mifflin Company, latest.

Or any other dictionary you prefer, like the Webster's cited in Section 20.2.3. In fact, having several different dictionaries in the department comes in handy. If one doesn't define a particular word, then another might.

Azar, Betty. *Understanding and Using English Grammar.* Englewood Cliffs, New Jersey: Prentice-Hall, 1981.

Not only the common but also some obscure grammar rules that help explain structures.

Brusaw, Charles T., Gerald Alred, and Walter Oliu. *Handbook of Technical Writing.* New York: St. Martin's Press, 1976.

Alphabetic reference including topic key and exhaustive index.

Carruth, Gorton, David H. Scott, and Beverly J. Yerge. *The Transcriber's Handbook with the Dictionary of Sound-Alike Words.* New York: John Wiley & Sons, 1984.

Grammar and punctuation rules and 2,750 sound-alikes.

Chicago Manual of Style, 13th edition. Chicago: University of Chicago Press, 1982.

The standard reference for writers, editors, and copy editors—first edition 1906. This thirteenth edition includes not only the current punctuation and format, but also current technology terms.

Fluegelman, Andrew, and Jeremy Joan Hewes. *Writing in the Computer Age: Word Processing Skills and Style for Every Writer.* New York: Anchor Books, 1983.

Choosing and using a word processing system.

Graham, Walter B. *Complete Guide to Pasteup.* Philadelphia: North American, 1975.

The skills required to prepare a master for print.

Hodges, John C. and Mary E. Whitten. *Harbrace College Handbook,* 7th edition. New York: Harcourt Brace Jovanovich, 1972.

Or any college handbook you prefer. Like having several different dictionaries, having several different handbooks among members of the department comes in handy. If one doesn't clarify a particular structure, then another might.

MIL-M-38784B, General Style and Format Requirements Technical Manuals, April 16, 1983.

Essential if your company writes publications to military spec, and a good reference in any case.

Strong, William. *Sentence Combining: A Composing Book*. New York: Random House, 1973.

Good for practice in the possibilities of sentence and paragraph structure.

Words into Type, 3rd edition. Englewood Cliffs, New Jersey: Prentice-Hall, 1974.

Based on studies by authorities who note some interesting details.

II. WRITING AND STYLE

The following books deal with the expository skills that every good writer needs.

Bernstein, Theodore M. *The Careful Writer: A Modern Guide to English Usage*. New York: Antheneum, 1977.

Just as the title says.

Follett, Wilson. *Modern American Usage*. New York: Hill and Wang, 1966.

The American version of Fowler.

Fowler, H. W. *Modern English Usage*, 2nd edition. New York: Oxford University Press, 1965.

A classic. Skim through it every once in a while.

Gunning, Robert. *The Techniques of Clear Writing*. New York: McGraw-Hill Book Company, 1968.

Probably the most popular "fog index" to calculate the grade level your writing is suited to.

Levin, Gerald. *Styles for Writing: A Brief Rhetoric*. New York: Harcourt Brace Jovanovich, 1972.

The diction, structure, and logic of composition.

Roget, Peter M. *Roget's International Thesaurus*, 4th edition. New York: Crowell, 1977.

To find the right word.

Strunk, William Jr. and E. B. White. *The Elements of Style*. New York: The Macmillan Company, 1959.

Maybe not everything you need to know, but much about grammar, punctuation, and style in the 78 paperback pages. (The third edition, 1979, is up to 85 pages.)

Urdang, Lawrence. *The Basic Book of Synonyms and Antonyms*. New York: Signet Books, 1978.

The right word—and its opposite.

III. REFERENCE AND RESEARCH

The following books deal with particular information that technical writers in electronics fields need.

Bartee, Thomas C. *Digital Computer Fundamentals*, 5th edition. New York: McGraw Hill Book Company, 1981.

Text on computer operations, structures, and organization.

Bell, C. Gordon and Allen Newell. *Computer Structures: Readings and Examples*. New York: McGraw-Hill Book Company, 1971.

A classic scholarly tome.

Bly, Robert W. *A Dictionary of Computer Words*. Wayne, Pennsylvania: Banbury Books, 1983.

Cartoons and analogies to help the beginner.

Burton, Phillip E. *A Dictionary of Microcomputing*. New York: Garland Publishing Company, 1976.

Comprehensive and clear definitions.

Encyclopedia of Computer Science. Anthony Ralston, Editor. New York: Van Nostrand Reinhold, 1976.

A wealth of information about electronic history, hardware, and software.

Freeman, Peter. *Software Systems Principles: A Survey*. Palo Alto, California: Science Research Associates, A Subsidiary of IBM, 1975.

A thorough discussion of computer structures, microprogramming, operating systems, languages, and case studies.

Kent, William. *Data and Reality: Basic Assumptions in Data Processing Reconsidered*. New York: North-Holland Publishing Company, 1978.

A marvelous trip back and forth between real and computer database worlds.

Mead, Carver and Lynn Conway. *Introduction to VLSI Systems*. Reading, Massachusetts: Addison-Wesley Publishing Company, 1980.

A dense, detailed, definitive text on silicon architecture and design using NMOS examples.

Meadows, A. J., M. Gordon, and A. Singleton. *Dictionary of New Information Technology*. New York: Nichols Publishing Company, 1982.

Words and acronyms spanning all communication devices.

Rosenberg, Jerry H. *Dictionary of Computers, Data Processing and Telecommunications*. New York: John Wiley & Sons, 1984.

Like the title, both book and print are large and comprehensive.

Sippl, Charles J. and Roger Sippl. *Computer Dictionary*, 3rd edition. Indianapolis: Howard W. Sams & Co., 1980.

Well cross-referenced words and acronyms.

Todd, Alden. *Finding Facts Fast*. Berkeley: Ten Speed Press, 1979.

A guide to reference sources.

IV. PLEASURE

The following books deal with technical material, and their authors raise technical writing to the stature of literature.

Bateson, Gregory. *Mind and Nature: A Necessary Unity*. New York: Bantam Books, 1979.

Provocative ideas and some wonderful patterns in which to arrange them.

Hofstadter, Douglas. *Godel, Escher, Bach: An Eternal Golden Braid*. New York: Vintage Books, 1979.

An amazing feat of synthesis. If you don't read anything else, read *Braid*.

Judson, Horace Freeland. *The Search for Solutions*. New York: Holt, Rinehart and Winston, 1980.

Science and technology as art and personalities.

Miller, Jonathan. *The Body in Question*. New York: Random House, 1978.

Technical metaphors adopted by medicine and psychology through the ages; Miller's "machine in the ghost."

Persig, Robert M. *Zen and the Art of Motorcycle Maintenance*, 2nd edition. New York: Bantam Books, 1980.

As Persig says, "The art of motorcycle maintenance is really a miniature study of the art of rationality itself." And he worked as a technical writer.

Zukav, Gary. *The Dancing Wu Li Masters: An Overview of the New Physics*. New York: Bantam Books, 1979.

Quantam mechanics for fun and great analogies and definitions for beginners.

GLOSSARY

The following terms refer to technical publications writing, editing, format, illustration, and production.

adjusted - A margin that has been changed from the standard and/or default.

alpha - Alphabetic.

appendix - A portion of a book following the last section and usually containing detailed information.

artwork - Figures and tables that are not entered with text, but have to be combined during production.

block diagram - An illustration using geometric shapes to show relationships in electronic and mechanical system.

boilerplate - Text and/or format used repeatedly with little or no variation.

book - A publication, finished or in process, whether it be a manual, guide, or overview.

callout - An identifying name, number, or letter assigned to a specific item in a figure.

camera ready copy - Text and artwork that is ready for reproduction.

character - An alpha or numeric representation, *A* through *Z* and *0* through *9*, or a symbol.

complement - A set of publications in support of the same product.

cutaway view - An illustration showing internal parts of an assembly that are not normally visible.

definition - A precise meaning of a word or phrase.

documentation - Research material, usually written by a product developer.

draft - A copy of the book before publication.

engineering drawing - An orthographic illustration of equipment in enough detail for the equipment to be manufactured.

exploded view - An illustration of an assembly that is separated, each part spread out in the proper sequence for reassembly.

figure - A graphic representation of information.

final copy - The edited, proofread, pasted up draft ready to become part of the master.

flowchart - An illustration showing passage of elements through a system.

font - A complete set of type of one size and typeface.

freeze date - The last day that copy can be changed. Also known as "drop dead date."

gloss - A brief explanatory phrase defining a word, phrase, or acronym.

glossary - A collection of glosses.

guide - A publication containing technical information, usually procedures or overviews.

hardcopy - Any information on paper.

header - The number and title of a section, subsection, sub-subsection or sub-sub-subsection.

index - An alphabetic list of significant subjects and the pages on which they are discussed in the publication.

justified - Flush with the margin.

key - A legend to identify the symbols in an illustration.

left-hand page - An even-numbered page of double-sided copy; also called verso.

logic diagram - A graphic representation of the interconnected decision-making elements of a system.

master - The copy that is reproduced for release.

mechanical edit - Reading a draft for references, notation, and consistency. Also known as copy edit.

one-sided - Pages that are blank on the back.

originator - The person who wrote the document or publication.

orphan - A lone first line from a paragraph or list at the bottom of a page.

new copy - Drafts incorporating editing changes.

pagination sheet - A form showing both sides of all pages to be printed on which you mark the page numbers (and "blank" if the page is to contain no text or figure).

parens - Parentheses (singular, **paren**).

part - A titled portion of a publication containing two or more sections.

preliminary - A publication that is incomplete, generally because of a significant amount of uncertain design details.

proofread - Reading a draft to ensure that changes marked on the original draft have been made in the new copy—and to find new typos.

publication - A book release by the publications department.

publication number - A unique number identifying a publication. Also known as a "document number" or "part number."

repro - Reproduction; that is, proofs to be photographically reproduced.

review - Reading a draft to correct technical inaccuracies.

right-hand page - An odd-numbered page; also called recto.

rub-ons - Sheets of symbols, numbers, and characters that are applied to paper by rubbing.

schematic - A graphic representation of a system or an assembly with connecting lines and symbols to show the flow of current.

section - The portions of a publication that carry a first-level header. Analagous to a chapter in a book. Also used generally to mean all section levels (1 through 4).

SOP - Standard operating procedure.

standard - A document that contains requirements for a company concern.

subsection - The portions of publication that carry a second-level header.

sub-subsection - The portions of a publication that carry a third-level header.

sub-sub-subsection - The portions of a publication that carry a fourth-level header.

table - A multicolumn representation of information.

technical edit - Reading a draft for content, organization, and style.

text - Paragraphs of prose.

two-sided - Pages that are not blank on either side.

typeface - A full range of type of the same design.

volume - A separately bound portion of publication.

widow - A lone last line from a paragraph or list at the top of a page.

wiring diagram - A schematic representation of wire connections, numbers, colors, and sizes.

INDEX

Abbreviations, 36–37
  common, 37
  period in, 27
Acronyms, 34–36
  communicating with, 124
  in glossary, 77, 78
  in index, 80
Active voice, 98
Adjectives, 86
  parallel, 118
  use in modification, 108–109
Adverbial conjunctions, 88
Adverbs, 86–87
  use in modification, 109
Alphabetization:
  dictionary method, 78, 80
  glossary, 78
  index, 80
  telephone book method, 78, 80
ALU, acronym, 35
Amounts, subject-verb agreement, 99
Analogies, comparative statements,
    139
Analysis, definition by, 138
Antecedents, 87
Antonyms, 139
Apostrophe, 30
Appendixes, 75
Appositives, 87
  as implicit definitions, 141
  use in modification, 109–110
Archive file, 199–203
  copies for, 200–201

distribution list/signoffs for, 201
  memos/notes for, 203
  reader comment reply form for, 202–203
Articles, 87
  definite, 87
  indefinite, 87
Audiences, 158–159
  applications/maintenance programmers,
    158
  customers, 159
  data entry personnel, 159
  maintenance personnel, 159
  preface and, 11
  professional nontechnical users, 159
  systems programmers, 158
  technical users, 158

Back cover, 15
Back matter, 14–16
  back cover, 15
  reader comment form, 14–15
  revision history, 14
Back-of-title page, 8–10
  copyright notice, 9–10
  disclaimer, 10
  reference to reader comment form,
    9
  revision of, 207–208
  revision notice, 9
  trademark notice, 9
Bit, acronym, 34
Block diagrams, 56
Bullets, 69

Callouts, 50
Cause and effect, 138
  definition by, 138
    if. . . then statements to define, 138
    statements of, 107
Cautions, 22
Change bars, 206
Clauses, 87–88
  dependent, 88
  independent, 88
  parallel, 119
  subordinate, 88
  transitional, 114
Climactic paragraphs, 112–113
Collective nouns, 91
  subject-verb agreement, 99
Colon, 29
Comma, 27–29
Command format, 41–44
  parallelism in, 120
  program externals, 41–43
  program internals, 42, 43
Command notation, 41, 42
Commands:
  computer, use of, 40
  defining, 138
Common nouns, 90
Comparative paragraphs, 112
Comparison, definition by, 138–139
Complex sentences, 105
Compound sentences, 105
Compound subjects, 99–100
Comprise, use of, 102
Computer communication, 39–46
  command format, 41–44
  commands, use of, 40
  error messages, 44–45
  instruction, use of, 40
  pseudocode, 45
  summary card, 46, 47
Conclusions, 115
Conjunctions, 88–89
  adverbial, 88
  compound subject, 99
  coordinating, 88, 106
  correlative, 88–89
  parallelism in, 119
  subordinating, 89
Consistency, 191–192
Content, 187–188
Contrast, definition by, 139
Cookbook approach, 147
Coordinating conjunctions, 88
Coordination, in sentences, 106–107
Copy editing, *see* Editing, mechanical edit

Copyright notice, 9–10
Correlative conjunctions, 88–89
Cost, of publications, 164
Count nouns, 90
Cover page for revision, 210–211
CPU, acronym, 35

Dash, 31–32
Declarative sentences, 95
Definite article, 87
Definitions, 135–141
  by analysis, 138
  by cause and effect, 138
  by comparison and contrast, 139
  explicit, 140
  of functions, 138
  in glossary, 77, 78
  implicit, 140–141
    by appositives, 141
    by nonrestrictive relative clauses, 141
  by location, 139
  by negation, 139
  of objects, 138, 139
  of operations, 138
  of procedures, 138
  of words, 136–137
Demonstrative pronouns, 95
Dependent clause, 88–89
Description, 143–150
  functional, 145–146
  physical, 143–145
  procedural, 146–150
    cookbook approach, 147
    tutorial approach, 147
Descriptive paragraphs, 112
Diagrams, block, 56
Diction, 123–124
  aesthetics, 134
  consistency in, 123–124
  economy, 132, 133
  gender, 131–133
  precision, 125–131
    confusing words, list of, 125–131
  terminology, 124–125
Dictionary method, 78, 80
Disclaimer, 10
DOS, acronym, 35
Double quotes, 33
Due to, use of, 102

Economy of diction, 132–133
Editing, 185–192
  copy/proofreader marks, 186, 187
  mechanical edit, 190–192
    for consistency, 191–192

for notation, 191
for references, 190–191
technical edit, 186–190
for content, 187–188
for organization, 188–189
for stylistics, 189–190
Ellipsis, 33–34
EMA, acronym, 35
Ergonomics, 39
Error messages, 44–45
parallelism in, 120–121
Exclamation point, 34
Exclamative sentences, 96
Expletives, 89
Explicit definitions, 140

Family tree, 53, 56
Figures, 49–61
block diagrams, 56, 57
flowcharts, 56, 58–61
hardware illustrations, 51
hierarchies, 53, 55–56
photographs, 61
rules for, 49–50
screens, 51–53
First person, 92–93
Flowcharts, 56, 58–61
logic flowcharts, 56
process flowcharts, 56
symbols, 56, 58
Footnotes, 23
Formalisms, 6
Front cover, 8
Frontispiece, 8
Front matter, 7–14
back-of-the-title page, 8–10
front cover, 8
frontispiece, 8
page numbers, 7
preface, 10–12
table of contents, 12–14
title page, 8
Functional description, 145–146
Functions, defining, 138
Future perfect progressive tense,
lie/lay, 103
Future perfect tense, 94, 96
Future tense, 96

Gender, 131, 132
General review form, 180–181
Gerundial phrases, 93
Gerunds, 89
Glossary, 77–78
alphabetization, 78

information included in, 77–78
long, 77
revision of, 208
short, 77
Got/gotten, use of, 102
Grammar, 85–121
adjectives, 86
adverbs, 86–87
antecedents, 87
appositives, 87
articles, 87
clauses, 87–88
conjunctions, 88
expletives, 89
gerunds, 89
infinitives, 89
modifiers, 89–90
nouns, 90
number, 91
object, 91
participles, 92
parts of speech, 92
person, 92–93
phrases, 93–94
prepositions, 94
progressive verb, 94
pronouns, 94–95
sentences, 95–96
subject, 96
tense, 96
verbals, 97
verbs, 97
voice, 97–98
Grammar usage rules, 98–104
pronoun reference, 101
subject-verb agreement, 98–101
verb forms, 101–104
Grammatical structures, 104–121
paragraphs, 112–113
parallelism, 116–121
sentences, 104–112
transitions, 113–115

Hardware illustrations, 51
Headers, 19–21
parallel headers, 116–117
section title, 19–20
spacing standards, 65
subsection title, 20
sub-subsection title, 20
sub-sub-subsection title, 20–21
Hierarchy charts, 53, 55–56
Hyphen, 31

IEEE, 36, 39

Imperative sentences, 95
Implicit definitions, 140–141
 appositives, 141
 nonrestrictive relative clauses, 141
Indefinite article, 87
Indefinite pronouns, 95
 subject-verb agreement, 100
Independent clause, 88, 89
Index, 79–81
 alphabetization, 80
 hardware books, 79
 revision of, 209
 software books, 79
Infinitive phrases, 93
Initialism, 35
Installation guides, 159
Instructions, 40
Interrogative sentences, 95
Interviews, 167–172
 defining terms in, 170
 guiding, 169–170
 props in, 167–168
 scheduling of, 167
 transcription of, 171–172
Intransitive verbs, 97
Introductions, 114–115
Irregular plurals, 91
Isometric illustrations, 51

Jargon, 124

K, acronym, 34

Lie/lay, use of, 102–103
Linking verbs, 97
Lists, 67–69
Location, definition by, 139
Logical components, 83–84
Logic flowchart, 56
Logos, 22
Loose sentences, 106

Mass nouns, 90–91
Master, 196
 publication, 196
 revision, 210–212
M-dash, 31–32
Mechanical edit, 190–192
Modification, 107–112
 adjectives, 108–109
 adverbs, 109
 appositives, 109–110
 participial phrases, 111–112
 placement of modifiers, 108
 prepositional phrases, 111

relative clauses, 110
Modifiers, 89–90
MOS/VLSI, acronym, 34
MVN, acronym, 35

Narrative paragraphs, 112
N-dash, 31–32
Negation, definition by, 139
Nonrestrictive relative clauses, 110
 as implicit definitions, 141
Notation, 25–48
 abbreviations, 36–37
 acronyms, 34–36
 computer communication, 39–46
 edit for, 191
 numbers, 37–39
 punctuation, 25–34
 spelling, 46, 48
 symbols, 39
Notes, 22
Noun phrases, 93–94
Nouns, 90–91
 appositives, 87
 collectives, 91
 common, 90
 count, 90
 gerunds, 89
 mass, 90–91
 number, 91
 proper, 90
Number, grammatical, 91
Numbers, 37–39

Object, 91–92
Operations:
 defining, 138
 describing, 145–146
Operator's guides, 159–160
Operator symbols, 39, 40
Organization:
 edit for, 188–189
 stated in preface, 11
Orthographic illustrations, 51

Page numbers, 7
 internal numbering, 17–18
 of parts, 18
 of sections, 17–18
Pages, revision of, 206, 207
Paragraphs, 112–113
 climactic, 112–113
 comparative, 112
 descriptive, 112
 narrative, 112
 topic sentence, 112

transitional, 114
Parallelism, 116–121
  clauses, 119
  conjunctions, 119
  headers, 116–117
  patterns of, 120–121
  phrases, 118–119
  point of view, 118
  words, 118
Parens, 32–33
Participial phrases, 93
  use in modification, 111
Participles, 92
Parts of speech, 92
Passive voice, 98
Past participles, 92
Past perfect progressive tense, 103
Past perfect tense, 96
Past tense, 96
Pattern of notation, 120–121
Period, 26–27
Periodic sentences, 106
Person, 92–93
Personal pronouns, 95
  pronoun reference, 101
Photographs, 61
Phrases, 93–94
  gerundial, 93
  infinitive, 93
  noun, 93–94
  parallel, 118–119
  participle, 93
  prepositional, 93
  transitional, 113, 114
  verb, 93–94
  verbal, 97
Physical objects:
  defining, 138, 139
  describing, 143–145
Planning a publication, 157–164
Plurals, 91
Point of view, parallel, 118
Preface, 10–12
  audience stated in, 11
  organization of publication in, 11
  purpose in, 10–11
  references in, 12
  revision of, 208
Prepositional phrases, 93
  use in modification, 111
Prepositions, 94
Present participles, 92
Present perfect tense, 96
Present progressive tense, *lie/lay*, 103
Present progressive verb, 94

Present tense, 96
Procedural description, 146–150
  cookbook approach, 147
  tutorial approach, 147
Procedures, defining, 138
Process flowchart, 56
Production, 193–196
  formatting pages, 194
  master, 196
  preparing final copy, 193–194
  standard components, preparation of, 195
Program externals, 41–43
Program internals, 43, 44
Program logic, pseudocode, 45
Progressive verbs, 94
Pronoun reference, 101
Pronouns, 94–95
  demonstrative, 95
  indefinite, 95
  number, 91
  personal, 95
  relative, 95
Proper nouns, 90
Proposal, 161
Pseudocode, 45
Publications, 159–160
  installation guides, 159
  operators guide, 159–160
  reference manuals, 159
  site planning guides, 160
  summary cards, 160
  system overviews, 160
Punctuation, 25–34
  apostrophe, 30
  colon, 29
  comma, 27–29
  dash, 31–32
  double quotes, 33
  ellipsis, 33–34
  exclamation point, 34
  hyphen, 31
  parens, 32–33
  period, 26–27
  semicolon, 29
  slash, 32
  underline, 33
Purpose, stated in preface, 10–11

Quotes, double, 33

RAM, acronym, 35
Reader comment form, 14–15
Reader comment page, 9
Reader comment reply form, 202–203
Reference manuals, 159

Reference to reader comment form, 9
References, 71–74
  edit for, 190–191
  stated in preface, 12
Relative clauses:
  as implicit definitions, 141
  restrictive/nonrestrictive, 110
  use in modification, 110
Relative pronouns, 95
Reprints, 213–215
Research, 165–173
  classes, 173
  conversations, 173
  documentation, 165–166
  hands-on, 172–173
  interviews, 167–172
Restrictive relative clauses, 110
Reviews, 179–184
  general review form, 180–181
  incorporating review material,
    184
  preparation for, 179–180
  return of, 182
  review approval form, 180
  technical review note, 182, 183
Revision history, 14, 208
Revision notice, 9
Revisions, 205–212
  of reprints, 213–215
  revision master, 210–212
  revision package, 206–210
Roman numerals:
  front matter, 7
  lists, 69
  parts, 18
Running feet, 18–19
Running heads, 18

SAC, acronym, 35
Schedule:
  of interviews, 167
  of publications, 162–163
Screens, illustrations for, 51–53
Second person, 93
Section headers, 19–20
Sections, 17–23
Semicolon, 29
Sentences, 95–96, 104–112
  complex, 105
  compound, 105
  compound complex, 105
  coordination in, 106–107
  declarative, 95
  exclamative, 96

imperative, 95
interrogative, 95
loose, 106
modification in, 107–112
periodic, 106
simple, 105
subjunctive, 96
subordination in, 106–107
transitional, 114
Signoffs, 201
Similies, 139
Simple sentences, 105
Site planning guides, 160
Sit/set, use of, 104
Slash, 32
Spelling, 46, 48
Stub, 63
Style, 151–154
  edit for, 189–190
  rules for, 151–152
Subject, 96
  compound, 99–100
Subject-verb agreement, 98–99
  amounts, 99
  collective nouns, 99
  compound subjects, 99–100
  indefinite pronouns, 100
  subjunctive sentence, 101
Subjunctive sentences, 96
Subordinate clause, 88
Subordinating conjunctions, 106
Subordination, in sentences, 106–107
Subsection title, of header, 20
Sub-subsection title, of header, 20
Sub-sub-subsection title, of header, 20–21
Summary card, 46, 47, 160
Symbols, 39
  flowchart symbols, 56
  operator symbols, 39, 40
System overviews, 160

Table of contents, 12–13
  relation to index, 81
  revision of, 208
  updating during writing, 176
Tables, 63–65
Tape recorder, use in interviews, 168–169
Technical edit, 186–190
Technical review note, 182, 183
Telephone book method, 78, 80
Tense, 96
Terminology, 124–125
Text, 21
Third person, 93

Title page, 8
    revision of, 207
Topic sentence, 112
Trademark notice, 9
Trademarks, 22
Transitions, 113–115
Transitive verbs, 97
Tutorial approach, 147
Typography, figures, 50

Underlines, 33

Verbals, 97
Verb forms, 102–104
    comprise, 102
    due to, 102
    got/gotten, 102
    lie/lay, 102–103
    sit/set, 104
Verb phrases, 93–94

Verbs, 97
    intransitive, 97
    linking, 97
    number, 91
    progressive, 94
    transitive, 97
Voice, 97–98

Warnings, 21
Words:
    defining, 135–136
    parts of speech, 92
    precision in use of, 125
    terminology, 124–125
    transitional, 113–114
Word use, *see* Diction
Writing process, 175–184
    organization, 175–177
    outlining, 176
    steps, 177–178